SEABUCKTHORN
Modern Cultivation Technologies

Author
Virendra Singh (India)

Co-Authors
Thomas S. C. Li (Canada), Lu Rongsen (China) and Y. Zubarev (Russia)

2008
DAYA PUBLISHING HOUSE
Delhi - 110 035

ISBN10 81-7035-521-4
ISBN13 978-81-7035-521-2
EAN 9788170355212

Published by : **Daya Publishing House**
1123/74, Deva Ram Park
Tri Nagar, Delhi - 110 035
Phone: 27383999
Fax: (011) 23244987
e-mail : dayabooks@vsnl.com
website : www.dayabooks.com

Showroom : 4760-61/23, Ansari Road, Darya Ganj,
New Delhi - 110 002
Phone: 23245578, 23244987

Laser Typesetting : **Classic Computer Services**
Delhi - 110 035

Printed at : **Chawla Offset Printers**
Delhi - 110 052

PRINTED IN INDIA

डा. एच. पी. सिंह
उप महानिदेशक
(बागवानी)
DR. H. P. SINGH
Deputy Director General
(Horticulture)

भारतीय कृषि अनुसंधान परिषद
कृषि अनुसंधान भवन-II
पूसा, नई दिल्ली 110 012
INDIAN COUNCIL OF AGRICULTURAL RESEARCH
KRISHI ANUSANDHAN BHAVAN-II
PUSA, NEW DELHI 110 012

Foreword

The dry temperate and cold deserts, spreading in over 75,000 sq. km area in the north-west Himalayas comprising Lahaul-Spiti and parts of Chamba and Kinnaur districts of Himachal Pradesh and Ladakh region, Jammu and Kashmir, are characterized by extreme cold conditions, high rates of soil erosion, shortage of fuel wood, timber and fodder, low productivity of agricultural lands and sparse vegetation. The past efforts of plantation by poplar and willow trees in the region have met only mixed success for the want of a multipurpose plant species that can simultaneously satisfy the long term conservation as well as economic needs of the people. Seabuckthorn *(Hippophae rhamnoides* L.) is multipurpose nitrogen fixing deciduous plant indigenous to the region, which meets above requirements.

Seabuckthorn, locally known as *"Chharma"* in Himachal Pradesh or *"Tsestalulu"* in Ladakh in Indian Himalayas is the highly nutritious soft fruit. Traditionally, this plant has been used by local medicine practitioners and village folks in treatment of various health ailments like digestive and skin diseases. Seabuckthorn is also a popular fodder, fuel wood and soil binder of cold deserts in Himalayas. Seabuckthorn has been harnessed in economic utilization and environmental conservation in countries like China, Russia and Germany for decades. In India, policy makers, scientists, industries, farmers and developmental agencies have recently joined hands and started plantation and utilization of this multipurpose plant for the production of various health protection products like multivitamin juices and cosmetics, helping thereby in upliftment of rural economy and ecological rehabilitation of fragile mountainous lands in the Himalayas.

The present book on *"Seabuckthorn: Modern Cultivation Technologies"* authored by Dr. Virendra Singh of CSK Himachal Pradesh Krishi Vishvavidyalaya (Agricultural University), Palampur and Prof. Thomas S.C. Li (Canada), Prof. Lu Rongsen (China) and Prof. Y. Zubarev (Russia) meets the urgent need for a practical, comprehensive technical guide for scientific cultivation and harvesting of seabuckthorn. I compliment the authors for bringing out this important and useful publication for the use of growers, scientists and students. I hope this publication will be of great help in commercialization of seabuckthorn for effective utilization of land profitably and improving the economic conditions of people who live in the region.

June 28, 2007

(H.P. Singh)
Deputy Director General (Hort.)

Ph.: (O) 25842068, (Resi.) 0120-2884682, (Mob.) 9871450730, Fax: 91-11-25841976
E-mail: hpsingh@icar.org.in

Preface

Seabuckthorn is hardly a house name in India; however, it has been widely used in traditional medicines by village folks and local doctors on Indo-China border high altitude areas of Himalayan states in India. It is also a popular fuelwood plant of the region. The region is known for extreme cold conditions and environmental problems high rates of soil erosion, causing siltation of rivers, consequently flood and related environmental and economic problems in mountainous and plain areas of north India. Willow, poplar and rubinia have been successfully planted in some pockets of the region, where as people in many areas remained unconcerned about this plantation programme because of no economic return. Of late, seabuckthorn, locally known as 'Chharma' in Indian Himalayas has been recognized as a solution to multitude of environmental and economic problems of the region. Fruit of seabuckthorn is very rich in vitamin C (300-2750 mg/100g), E, beta-carotenes, flavonoids and sterols etc., which have potential in health food, cosmetic and drug industries of the country. During the last one decade, after learning of success stories of seabuckthorn in Russia, China, Germany and Finland etc., CSK Himachal Pradesh Agricultural University, Palampur created awareness about the potential of seabuckthorn in environmental conservation and economic development of the region. Efforts of this university were later joined by Defence Research and Development Organization (DRDO) and other research and developmental organizations of the country, who joined hands, adopted a multi-institutional and multidisciplinary approach in development of appropriate technologies for seabuckthorn development in the country. Consequently, cultivation, processing and marketing of various seabuckthorn health products have started in the country and seabuckthorn is now becoming a popular plant among policy makers, scientists, farmers, environmentalists and health industries.

Managed scientifically for commercial production, seabuckthorn provides an excellent opportunity in establishing a value added crop on marginal lands. These shrubs, planted closely, will produce high yields of tightly clustered berries, with bumping crop in 5-6 years and remain productive for 30 years or even more, by proper pruning.

As seabuckthorn development in India and most of other countries is in early stage, this small book on seabuckthorn has been prepared with an intention to give boost to commercial cultivation of seabuckthorn to meet the increasing demand of raw material of seabuckthorn not only in Indian market and but also global market. Authors of the book have compiled the all globally available latest information on the modern technologies on the propagation, plantation, management and fruit harvesting of seabuckthorn, which will help in the large scale commercial cultivation and utilization of this multipurpose plant for the production of various health products for the benefit of common man, improvement of rural economy and of course, environmental conservation of fragile mountainous lands. The book has been particularly prepared for commercial growers of seabuckthorn; however, it will also prove useful for academicians, environmentalists, farmers, foresters etc. and those interested in seabuckthorn.

Authors

Corresponding Author

Dr. Virendra Singh
CSK Himachal Pradesh Agricultural University,
Hill Agricultural Research and Extension Centre,
Bajaura – 175125 (Kullu), India
(E-mail: virendrasingh2003@yahoo.com)

Contents

Chapter 1
Introduction

Seabuckthorn (*Hippophae rhamnoides* L.) is a 2-5 m tall thorny, nitrogen fixing, deciduous shrub or small tree, a unique and valuable multipurpose plant presently being grown in cold regions of over 40 countries, including India, China, Russia, Europe and Canada. In India, seabuckthorn grows naturally in high altitude regions of Himalayan states of Himachal Pradesh, Ladakh (J&K), Uttrakhand and northeast India (Sikkim and Arunachal Pradesh) (Singh, 2003). Fruit of seabuckthorn is rich in vitamin C (300-2750 mg/100g), which is 4-100 times richer than many other horticultural crops and vegetables. Fruit oil (4 per cent) has anti-bacterial, anti-inflammatory, tissue regenerating and anti-tumor properties. Because of its richness in anti-oxidants like vitamin E, carotenoids, flavonoids and sterols and unique unsaturated fatty acids, its oil has found application in treatment of diseases like cardiovascular, cancer, healing of wound and gastric ulcer and skin diseases and utilization in development of cosmetics (Yang and Kallio, 2006). Plant is very hardy and can withstand –40°C in winter and naturally adapoted to low rainfall (50-500 mm yr^1) and high snowfall areas of the Asia (upto 5300 m asl) and Europe. It grows naturally on sun loving slopes and sandy riversides. It requires irrigation for cultivation in areas with rainfall less than 300-400 mm. In India, there are about 12,000 ha land area under seabuckthorn, mostly in Ladakh. Fruit and leaves are rich in variety of vitamins and other bioactive substances, having medicinal properties. It has been recognized as "a food or food ingredient, which has health benefits, including the prevention and treatment of diseases. Medicinal values of seabuckthorn were discovered as early as 8th century, as mentioned in Tibetan Medical Classic *"rGyud Bzi"* After discovery of vitamin values of seabuckthorn by Russians in 1940s, plant was well researched and many health products have been developed in Russia and China. Now plant is being utilized in health food, cosmetic and drug industries in other countries also like India, Canada, Germany, Finland and Sweden. It has also been introduced in USA, Bolvia, Korea and Japan. Plant is also a useful fodder, fuelwood and soil binder. Its nitrogen fixing roots take roots and multiply rapidly even in poor and dry sites. Seabuckthorn is an ideal plant for soil erosion control, land reclamation, wildlife habitat conservation and soil conservation and improvement in fragile ecosystems especially mountainous lands. Seabuckthorn has been widely planted to control soil erosion in mountainous lands in China. China has the maximum resources of seabuckthorn, followed by Russia, Mongolia, Kazhakstan and India. Other countries particularly European countries have very limited resources of seabuckthorn.

Plant start bearing fruit, a berry, after 3 years of plantation and with application of better cultivars and management techniques, it remain productive for next 25-30 years, producing an average of 15 tons of fruits per ha per year. Till few years back, demand for seabuckthorn raw material in industries was very limited and people around the world were meeting the requirement of raw material from the wild sources. However, due to standardization of health products of seabuckthorn along with improvement of processing technologies, the demand for quality raw material of seabuckthorn has increased rapidly. Commercial cultivation of seabuckthorn can only be economically viable with better and characterized plant material and improved agrotechniques for high productivity of seabuckthorn. Further, growers need to be assisted in increasing the fruit yield by providing them appropriate harvesting machines.

Russians as early as 60 years back were first to start the breeding work, which started with surveys and selections of promising forms of seabuckthorn, as wild forms were highly thorny and low yielding (1-2 tons/ha). M.A. Lisavenko and his team, who had started work as early as 1930s, started pioneering work. Many thornless, high yielding, rich in fruit oil and vitamins etc. with large fruit size and high productivity (15-25 tons/ha) were released over the last two decades (Kalinina, 1988), through selection, hybridization and even mutagenesis. German, Finnish, Canadian, Chinese and Mongolian experts started work on genetic improvement of seabuckthorn. In India, scientists of CSK Himachal Pradesh Agricultural University, Palmpur, Field research Laboratory (DRDO), Leh and G.B. Pant Institute of Himalayan Environment and Development, Srinagar (Uttrakhand) have been working on seabuckthorn and high yielding cultivars of *H. rhamnoides* ssp. *turkestanica* and *H. salicifolia* are under evaluation.

Propagation of seabuckthorn from seeds is successful, however, there is no control on the ratio of male and female plants. The method has been considered for introduction of exotic forms and raising large nurseries for environmental conservation. For meeting the demand of orchards, large nurseries were raised from hardwood cuttings of seabuckthorn in Russia, China and India. However, method is also prone to environmental stress. During the last one decade, workers standardized the propagation of seabuckthorn from softwood cuttings under control conditions of mist system. Growth hormone NAA and IBA have been found quite useful in rooting of soft woods of seabuckthorn. The method has proved very efficient and economically less expensive. There are some progresses on standardization of micropropagation of seabuckthorn in India, Canada, China and Finland. While multiple shoots formation has gone upto 15 shoots/explant (Singh and Gupta, 2007), many experts are working on improvement of rooting rates and hardening of explants under greenhouse and field conditions.

While plantation of seabuckthorn on plain land is quite easy, Chinese experts developed methods for plantation of seabuckthorn on fragile slopes, even without proper irrigation, which has helped in both, resource regeneration of seabuckthorn and environmental conservation. Main challenge has been standardization of management practices, *e.g.* appropriate spacing, weeding, pruning regime, nutritional requirement, identification and control of diseases and pests. There is huge information on some aspects like spacing of an orchard of seabuckthorn. In British Columbia, seabuckthorn plantation with 2500 trees/ha at a male and female ratio of 1:8, produced 20-25 tons fruits/ha. Some of progress has been made in standardization of agrotechniques of seabuckthorn. Still much need to be done on standardization of weeding, pruning regimes and nutritional needs of the various subspecies and cultivars of seabuckthorn. *Fusarium* wilt a serious disease of seabuckthorn and there is no control method. Pests like *Holcocerus hippophaecolus* have also recently spread in some parts of China and control measures are being developed to control them (Luo *et al.*, 2007). There is a need to standardize the organic and biological control measures and develop diseases and pest resistant cultivars of seabuckthorn. Still, there are huge technological gaps in seabuckthorn cultivation programme in India and efficient orchard management practices need to be developed for forms of seabuckthorn under various soil and climatic conditions.

Harvesting of seabuckthorn fruit manually has proved very difficult as the plant is thorny and labour expensive. Therefore, during the last one decade or so, a number of small tools have been globally developed, which have made fruit collection more easy and more productive. Good efforts have been made to develop the machines in Russia, Germany, Sweden and Canada, who have fitted shaker on a tractor, which shake or vibrates the plant and ripe fruits are separated from the plant, increasing fruit harvest hundred times faster than manually. However, there is a less fruit collection in case of varieties with small fruit and peduncle, where there has been more damage to the plant and lot of mixture of leaves and shoots in the fruit collected. Recently, harvesting machine has been developed in Germany for small fruit also, however, it is still very expensive especially for developing countries.

During the last one decade, huge research work has been done on improvement of cultivation technologies of seabuckthorn in countries like Russia, China, Germany, Finland and Canada etc., which can be shared with growers in India and globally, which are in early stage of seabuckthorn development. Therefore, we have made an effort to provide the required information on the application of appropriate technologies generated globally for seabuckthorn resource development, which may certainly be of great help in the commercial cultivation of seabuckthorn in India and across the world. Besides, helping in economic upliftmenmt of rural economy particularly in developing countries, the commercial cultivation of this multipurpose plant, both on marginal and agricultural lands is likely to improve the conservation of fragile mountainous lands and wild life and over all ecosystem of the seabuckthorn growing regions of the world.

Chapter 2
Plant Characteristics

Seabuckthorn is a pioneer species, colonizing the open sites like abandoned agricultural lands, infertile wastelands, riversides, hill tops, slopes and rocky lands, as pure stands or mixed stands in association with other plant species in cold regions of Asia and Europe (Lu Rongsen, 1992). Traditional name of seabuckthorn varies from country to country, place to place or language to language or even community to community. Seabuckthorn is known as Chharma or Tsestallulu in India, Shaji in China, Tarbu in Tibet (China), Oblepikha in Russia, Sanddorn in Germany, Duindoorn in Netherland, Olivello Spinoso in Italy, Tyrni in Finland, Argousier in France, Espino Armarillo in Spain, Finbar in Sweden, Tindved in Denmark, Rokitnik in Poland and Yashildoo Chatsagana in Mongolia.

It succeds in most soils, including poor sites, so long as they are not too dry. It also grows in wet sandy soils of river sides with running water. The standing water kills the plant. Seabuckthorn is a perennial, woody, nitrogen fixing and thorny, deciduous shrub of 1-4 m height, covered with silvery scales. The plant stem or branches have rough brown bark. Male and female plants are separate. Plant shows shrubby nature under natural conditions of growth, where many plants grow in dense population. The age of the plant is about 30-60 years. *H. rhamnoides* is a shrub species of 2-4 m height. However, tall trees have also been found growing, where there is lot of distance between the plant and there is plenty moisture, *e.g.* river sides or farmers' fields, where it can reach to 6-10 m in height. A very tall tree of 16 m height with stem girth of 5.3 m has been found in Muli County, Sichuan, China. One species of seabuckthorn *H. salicififolia* generally has trees of medium height of 6-9 m, whereas *H. tibetana* is an under shrub of 0.2-0.6 m, growing in alpine zone of Himalayas (Singh *et al.*, 1995). Following are some characteristics of seabuckthorn (Lu Rongsen, 1992):

Leaves

Alternate and willow like leaves of seabuckthorn are borne 7-10 days after flower buds. Length of the leaves is usually 3 cm to 8 cm and width 0.4 cm to 1.2 cm. Leaves are narrow, linear, lanceolate and covered with silvery stellate scales on the upper side, which reflect the sun rays and reduce the evaporation from the plant. Leaves are rich in vitamin C (150-250 mg/100g) and flavonoids, therefore rich in anti-oxidants and have application in tea, cosmetics, cardiovascular diseases and wound healing. Leaves as a tea crop, should be harvested, from the male plants only, during late June and mid July, so as not to interfere with the fruit set in the female plants.

Buds

The vegetative buds develop generally on male plants, which are not bearing fruit. In fruiting plants, the mixed buds are developed. In male plants, the buds are larger, more protruding and contain 6-8 covering scales. In female plants, the buds are smaller, more elongated, sessile, and have only two covering scales.

Flowers

Seabuckthorn is a diocious plant, where, male and female plants are separate. The flower buds are borne every year during active growth season of summer or autumn, along with vegetative buds and open in spring during the next year. Male buds are larger than the female buds, with 6-8 covering scales. The smaller female buds are more elongated and have 2 covering scales. Male and female buds open at the same time about 7 days before leaf buds. Very small yellowish pistolate flowers appear, usually before the leaves, late April to mid May. Both male and female flowers are very small. The female flower is without petals and it consists of a pistil, a hypanthium and 2-lobed perianth and found in small racemes in the leaf axils (Rousi, 1971). The male flower has longer oblong perianth leaves and contains four stamens of same length and it also has no petals. Male and female flowers have no nectarines, therefore, seabuckthorn is wind pollinated. The anther splits and pollens are blown out in large numbers to female flowers by wind, when the atmospheric temperature is in the range of 6-10°C. Female flowers are receptive to pollens for 10 days. Some insects like honey bees visit to male flowers to eat pollens, however, they do not visit female flowers. The female flower bud contains generally one flower, rarely 2-3 flowers, whereas, male flower bud contains 4-6 flowers. Vegetative shoots are formed from latent buds on 2-4 year old woody branch. Mixed shoots are borne from buds on 1-yr old woody branch. Flowering shoots are borne at the base of vegetative and mixed shoots.

Male and female plants can not be differentiated unless the female plants bear fruits, which become conspicuous in July month and mature in late August or early September.

Fruits

Seabuckthorn fruits are of pea size, often less than 1 cm in diameter, varying in weight of 100 fruits from 15 to 25 g. Some Russian forms are of higher weight (40-90 g/100) (Kalinina, 1988). Fruits are of attractive colours of red, reddish-orange or yellow, therefore this species was planted as a garden plant for beautification in Europe. Fruit shape varies to ovate, spherical, cylindrical, elliptic etc. Seabuckthorn fruit is a typically berry, which is developed from an ovary and expanded calyx tube. Therefore, the fruit is made from unsplit, fleshy, expanded calyx tube and an ovary. The period from flowering to fruit maturation is 3 to 4 months. Most of the fruits are borne on 2-year-old branch. The young fruit is green and hard, which become soft on maturation in late August or early September. Mature fruits may remain attached to the female plants for several months, unless they fall by wind or snow or eaten by birds. They are generally harvested from late August to early October. Later in winter they become over mature and shrink and often eaten by wild animals especially by birds. One female plant generally bears 0.2 kg in an young plant to 3 kg in a mature plant of 10 years or so. One can harvest 2-5 tons of fresh fruits from wild forest of seabuckthorn. Improved varieties have been reported to produce 8-18 kg of fruits per plant or 10-40 tons per ha. Female plants bear fruits after 4-5 years in plants raised from seeds and 3-4 years in plants raised from stem cuttings.

Seeds

Each fruit has a single seed surrounded by a fleshy hypanthium (edible fleshy pulp). Seeds make 5-15 percent of fruit weight and they are of 3-7 mm in length and 2-4 mm in width and light to dark brown in colour. The colour of seeds of *H. tibetana* are comparatively larger and dull white in colour (Singh *et al.,* 1995). The seed is surrounded by a parchment like ovarian wall. There are about 1,25,000 seeds per kg of seabuckthorn seeds.

Thorns

Seabuckthorn is a highly thorny plant. Thorns are borne on the branches and main stem along with fruits and leaves. There are generally 3-5 thorns per 10 cm of 2-year-old branch of *H. rhamnoides*. *H. salicifolia* is a mild thorny tree, whereas thorns are borne on shoot tips in *H. tibetana.* The length of the thorn may vary from 2 to 5 cm. Mild thorny or thornless varieties of seabuckthorn have been produced through natural selection or mutagenesis in Russia. Thorns create some problem in fruit and leaves harvesting by hands, however, new methods have been devised in their easy harvesting by improved hand tools and machines.

Roots

Seabuckthorn has well developed strong root system, which tap both surface as well as underground moisture. The root system consists of light coloured, thick, rope like primary root with many adventitious rootlets. Old plants have been found to have horizontal roots extending upto 6-10 m, most of them lying subsurface (10-30 cm soil depth) and 1-3 m vertical roots, which help in exploitation of soil moisture and nutrients and establish in infertile and difficult sites in a better way than many other plants. Seabuckthorn horizontal coarse root give birth upwards to several root turion seedlings or sister plants of several generations, also known as root suckers, making a network of complex root and a dense stand or forest, which hold soil particles even on fragile steep slope. It is because of fast regeneration of seabuckthorn plants and a strong network of roots, therefore, seabuckthorn afforestation has been used on large scale in controlling of soil erosion, protecting river banks, preventing of floods in China.

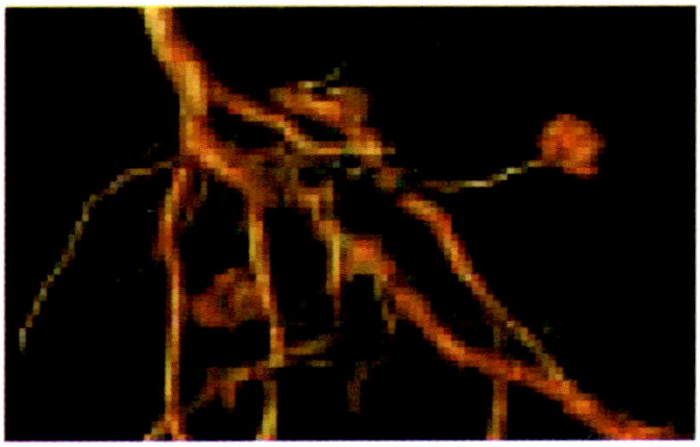

Plate 1: Nitrogen Fixing Root Nodules of Seabuckthorn

Root Nodules

A symbiotic nitrogen fixing mycorrhizal bacteria *Frankia* has been found, which make nodules in the roots of seabuckthorn (Gatner and Gardener, 1970). Seabuckthorn root nodules have been estimated to fix about 180 kg of atmospheric nitrogen per ha per year, which is as good as soybean. Root nodules also help the plant to transform the non-dissolving soil material to dissolving state. Seabuckthorn, when planted in a sand dune, increased nitrogen in soil by 1.5 times within 3 years of plantation.

Chapter 3
Biochemical Characteristics

Fruit of seabuckthorn is very rich in variety of vitamins and other bioactive compounds with nutritional and medicinal properties (In: Li and Beveridge, 2003; Singh 2006; Yang and Kallio, 2006ab).

Fruit

Weight of 100 fruits may vary from 15 to 70 g, although in some cases, it may be as low as 8 g or as high as 110g. The fruit may contain 60-85 per cent juice and it may yield about 65 per cent juice by centrifugal method (Heilscher and Lorber, 1996). Each fruit contains a seed, weighing on average 15-20 mg, containing 11 per cent moisture and 8-18 per cent oil.

Dry Matter

In the ripe seabuckthorn fruits, dry matter is represented by fat and non-lipid fractions. Generally dry matter constitutes about 15 per cent of the total weight of the fruit. It is known that small-fruited seabuckthorn contains more dry substances than large-fruited one.

Soluble Solids

Soluble solids represent an important fraction of the seabuckthorn juice, because of the high concentration of organic acids typical of its berry. Chinese seabuckthorn berries possess the highest content, varying from a minimum of 5.6 to a maximum of 22.7 °Brix.

Proteins and Amino Acids

Total protein level in fresh fruits of seabuckthorn is in the range of 2.1-3.4 per cent. In seeds, it may be between 18-33 per cent. Protein content in the pulp varies from 0.79-1.64 per cent, that is why pulp or juice seems cloudy or opalescent product, which provides a stable turbidity to the juice. The majority of seabuckthorn proteins are well-ingested albumins and globulins. As studied in Russian forms, globulins (53.7-56.0 per cent) and albumins (33.1-38.4 per cent) are important proteins, which contain a large number of free amino acids, among them aspartic acid is quantitatively most important.

Pectin

Pectin content of the seabuckthorn fruits is low. Pectin value was estimated in Siberian seabuckthorn (var. Katun) ranging from 0.2 to 1.2 per cent.

Sugars

Although, seabuckthorn berries are not considered rich in sugars, however, sugar is an important ingredient of seabuckthorn fruit, as it plays a useful role in determining the sweetness of its juice and in fact the sugar:acid ratio has been reported to constitute the major promoter of taste of seabuckthorn fruit juice. The average content of sugar in fruits is 2.00-3.26 per cent and in the sweetest Russian forms, it can go up to 7.0 per cent. The sugar is composed of glucose (1.3-1.8 per cent), fructose (0.7-2.3 per cent), and saccharose (0.07-0.30 per cent). There are minor amounts of xylose, mannitol, sorbitol and xylitol.

Organic Acids

Organic acids and sugars are the major portion of the soluble solid fraction of the fruit pulp of seabuckthorn fruit. Fruit juice of seabuckthorn is quite rich in organic acid, pH of juice being near to 2.7. Chinese species showed the highest content of organic acid (4.1 to 9.1 per cent) and 2.1-3.2 per cent in some Russian forms. About 90 per cent of the total acidity is represented by malic and quinic acids in Chinese, Russian and Finnish berries, malic acid being a major constituent. Presence of vitamin C organic acid and tannic acid in the fruit of seabuckthorn make it an ideal sources for the production of several beverages particularly health protection juices.

Vitamin C

Seabuckthorn is famous for very high content of vitamin C (100-2750 mg/100g), which are 4 -100 times higher than any vegetable and fruit. Generally, Chinese seabuckthorn (ssp. *sinensis*)-360-2500 mg and *H. salicifolia* (2750 mg) in Indian Himalayas, are richest in vitamin C, whereas Russian and European forms are low to medium. It is anti-oxidant, scavenger of free radicals, inhibits the formation of potentially carcinogenic N-nitroso compounds and thus offers protection against stomach cancer. Ascorbic acid also plays a critical role in wound repair and healing/regeneration process.

Oil

Oil of seabuckthorn berries is the most valuable product of this plant, as it possesses anti-oxidant, wound healing, anti-ulcer, anti-tumor and curing cardiovascular etc. properties. Generally, the oil content of seabuckthorn fruit is low (about 4 per cent), whereas as ssp. *turkestanica* in western Pamirs, Tajikistan is quite rich in oil of fresh fruits (6.8-13.7 per cent). Russian and central Asian forms are known to be rich in oil, where the oil content on the average does not fall below 6.0-6.6 per cent. In Lahaul valley, Indian Himalayas, total oil in fresh fruits varied from 2.9-4.6 per cent in *H. rhamnoides* ssp. *turkestanica* and much lower of 2 per cent in *H. salicifolia*. The total oil content varies from 2-4 per cent in pulp to 8-16 per cent in seed.

Fatty Acids

Unsaturated fatty acids make about 85 per cent of total oil. The human body absolutely requires the polyunsaturated EFAs α-linoleic acid (omega-6 fats) and alpha-linolenic acid (omega-3 fats). Both fatty acids repair the cell membrane after oxidation due to attack of free radicals. Seabuckthorn seed oil is very high in two essential fatty acids, Linoleic acid (30-40 per cent) and α-Linolenic acid (20-

35 per cent). The dominating fatty acids in the soft parts of the fruit are palmitoleic acid (16-54 per cent). Generally less than 14 per cent of Linoleic acid and less than 3 per cent of α-Linolenic acid are usually found in pulp oil. Palmitoleic acid has attracted an increasing interest due to its possible effects on many physiological processes including cholesterol and triglyceride lowering and stroke reducing effects. Palmitoleic acid is a principal constituent of skin fat and the extract is recommended for skin softening and anti-wrinkle products.

Vitamin E

Important antioxidant function of vitamin E, also known as tocopherols, appears to be the inhibition of lipid peroxidation, scavenging free radicals. Low intakes of vitamin E and other anti-oxidants results into certain types of cancer and atherosclerosis. Both pulp and seed oils of seabuckthorn are rich in vitamin E, much higher than other nutrient oils. Chinese experts have found the highest vitamin E content in seed oil of *H. rhamnoides* subsp. *turkestanica* (159 mg/100 g) and lowest in *H. solicifolia* (46.9 mg/100 g oil) and). The highest vitamin E content in pulp oil was found in *H. rhamnoides* subsp. *sinensis* (248 mg/100 g oil).

Vitamin K

Vitamin K promotes normal coagulation of blood during the injuries of blood vessels, and its content varies from 0.65-1.3 mg/100gm of fresh fruit, 59-64 mg/100g in pulp oil to 110-230 mg/100g in seed oil, which is more than many horticultural crops.

Carotenoids

Various colours of ripe berries of seabuckthorn, ranging from yellow to bright red are related to occurrence of carotenoids. Carotenoids are linked with a decreased risk of heart disease, cancer, and degenerative eye diseases like macular degeneration and cataracts. Beta-carotene, the most dominant carotenoid in seabuckthorn, intake is associated with reduced risk of breast, stomach, esophageal, and pancreatic cancers. Total carotenoid content in seabuckthorn fresh fruit varies generally from 1 mg to 120 mg/100g, whereas content of β-carotene varies from 0.2 to 17 mg/100g. Red and orange-red fruits are richer in carotenoids as compared with the less intensely coloured fruits like yellow and orange-yellow.

Different growth conditions influence the carotenoid content in the soft parts of the berries. Carotenoid content has been found increasing with the maturation of fruit. Soft part (pulp) oil of ssp. *sinensis* growing in Shanxi province of China had a maximum carotenoid level of 2140 mg/100g and a minimum value of 2.1 mg/100g in north Caucasus. It is also established that carotenoids of seabuckthorn fruits consist of α-carotene, β-carotene, lycopene and zeaxanthin. Many studies have found β-carotene to be a major carotenoid, making 15-55 per cent of total carotenoids, depending on the place.

Table 1: Chemical Composition of Fruit of Some Important Species and Subspecies of Seabuckthorn (*Hippophae* L.)

Parameter	*Range (Mean)*	*Species/ Subspecies*	*Reference*
Fruit weight	15-27	*turkestanica*	Singh *et al.* (1995)
(g/100)	30.5	*H. salicifolia*	Singh *et al.* (1995)
	30-90	*mongolica*	Kalinina (1988)
	17.5-27.5	*sinensis*	Lu Rongsen (2006)
Fruit moisture	72-75 (74)	*sinensis*	Ma *et al.* (1989)
(%)	74-85 (82)	*sinensis*	Li *et al.* (2002)
	95.0-81.8 (88.3)	*sinensis*	Lu Rongsen (2006)
Vitamin C	160-340	*H. rhamnoides*	Albrecht (1990)
(mg/100g)	165-293	*rhamnoides*	Rousi and Aulin (1977)
	1294	*H. rhamnoides*	Eliseev (1976)
	867.5-1351.5 (1109.5)	*sinensis*	Lu Rongsen, (2003)
	219-642	*turkestanica*	Singh *et al.* (1995)
	2740	*H. salicifolia*	Singh and Sawhney (2005)
	138–1400 (769)	*turkestanica*	Dwivedi *et al.* (2006)
	360-2500	*sinensis*	Li and Schroeder (1996)
Total fruit oil (%)	6.2-12.6	*turkestanica*	Korzinnikov *et al.* (1981)
	2.5-4.9	*mongolica*	Shirko and Radyuk (1989)
	1.5-1.6	*sinensis*	Wei and Guo (1996)
	2.9-4.6	*turkestanica*	Singh and Singh (2004)
	2.0	*salicifolia*	Singh and Singh (2004)

Contd...

Table 1–Contd...

Parameter	*Range (Mean)*	*Species/ Subspecies*	*Reference*
Juice oil (%)	0.3-1.4 (0.9)	*sinensis*	W. Zhang *et al.* (1989)
	1.8-2.9 (2.0)	*sinensis*	Ma *et al.* (1989)
Refractive index	1.349-1.356 (1.353)	*sinensis*	Tong *et al.* (1989)
Specific gravity	1.027-1.045 (1.037)	*sinensis*	Tong *et al.* (1989)
Soluble sugars (EBrix)	6.4-12.7 (9)	*sinensis*	Ma *et al.* (1989)
	10.2-22.7 (16.0)	*sinensis*	W. Zhang *et al.* (1989)
	10.8-15.6 (13.5)	*sinensis*	Tong *et al.* (1989)
	8.5-9.0 (8.8)	*sinensis*	Lu Rongsen (2006)
Glucose of total sugars (%)	49.5-62.1 (54.2)	*sinensis*	Ma *et al.* (1989)
Fructose of total sugars (%)	37.3-50.4 (45.4)	*sinensis*	Ma *et al.* (1989)
Organic acid (%)	4.6-7.3 (6.0)	*sinensis*	W. Zhang *et al.* (1989)
	3.5-4.4 (4.0)	*sinensis*	Ma *et al.* (1989)
	5.2-6.3 (5.7)	*sinensis*	Lu Rongsen (2006)

Flavonoids

Seabuckthorn fruit and leaves are very rich source of flavonoids. Flavonoids are found in all parts of seabuckthorn, *i.e.* leaves (3.8-4.0 per cent), fruits, juice and seeds. Russians have estimated 420-552 mg/100g flavonoids fresh fruits of various seabuckthorn populations. Seabuckthorn growing in west Pamirs had flavonoids 310-1238 mg/100g dry wt in leaves and 168-859 mg/100g in crude fruits. Studies found that the juice and dried fruit residue contained flavonoid of 0.2 per cent and 0.55 per cent, respectively in seabuckthorn from western Sichuan, China. The main flavonoids identified in seabuckthorn are leucocyanidin, catechin, flavonol and trace flavanone. From flavonol, the isorhamnetin, quassin and camellin could be isolated. They have been found to possess very strong anti-oxidant activity. It has been found that flavonoids improve the immunity of the body, lower the osmosis of the capillary wall and prevent oxidation of vitamin C. Flavonoids have been found in controlling arteriosclerosis, reducing cholesterol level, turning hyperthyroidism into euthyroidism and eliminating inflammation. They have also been found effective against tumor and radiation damage.

Sterols

Sterols constitute the main portion of unsaponificable matters. All sterols in seabuckthorn oil belong the following 4 series, *i.e.*, ergosterols, stigmsterol, lanosterols and amyrins. Due to their structural similarity to cholesterol, plant sterols are well studied for their cholesterol absorption inhibition properties. In addition to their cholesterol lowering property, plant sterols may possess anti-cancer, anti-atherosclerosis, anti-inflammation and anti-oxidation activities. The amount of sterols, in the fruit pulp of studied forms ranged between 0.16 and 0.76 per cent, but in the seeds, it ranged from 0.19 to 0.96 per cent. It was found that content of sterol in seabuckthorn oil is about 10 times higher than other oils. Total sterol content in the pulp oil (soft part) of seabuckthorn fruit ranged from 1 per cent to 3 per cent. Juice oil, processed by centrifugation of pressed juice of subsp. *sinensis*, had 720 mg/100g sterols.

Folate

Folate is a water-soluble vitamin B known to have several benefits to human health, such as prevention of neural tube defect in babies, an action against cardiovascular diseases caused by elevated plasma homocysteine and certain forms of cancer. Seabuckthorn fruits have been found to be a rich source of folate (39 µg/100 g fresh weight).

Betain

Seabuckthorn accumulates betain an anti-ulcer compound in high amount. It varied from 19.9 to 190 mg/100g in seabuckthorn cultivars growing at Urals Curative Plants Garden, Russia. Betain quantity in seabuckthorn fruits varied from 512 to 897 mg per cent in Altay cultivars and from 728 to 1389 mg per cent in East Sayan forms.

5-hydroxytryptamine (5-HT)

Of the chemical neurotransmitter substances, serotonin is perhaps the most implicated in the etiology or treatment of various disorders, particularly those of the central nervous system, including anxiety, depression, obsessive-compulsive disorder, schizophrenia, stroke, obesity, pain, hypertension, vascular disorders, migraine and nausea The peel of stem and fruit of seabuckthorn contains serotonin. In Russian forms, experts estimated 1.1-2.6 mg/100g serotonin in seabuckthorn fruit. 5-hydroxytryptamine (5-HT) isolated from seabuckthorn bark inhibited tumor growth.

Tannins

Seabuckthorn leaves have been found to contain high content of polyphenols, including tannins (10-12 per cent). Therefore, seabuckthorn plant leaves have been proposed as the prospective source for dyeing and tanning substances. Fruits, pulp and juice were found to be poor in tannins (0.02, 0.02 and 0.004 per cent, correspondingly). Seabuckthorn tannins are important source of anti-viral drugs.

Metallothionein

Metallothionein acts as detoxying agenct for heavy metals and as free radical scanvengor for most toxic radical, hydroxyl radical (HO^-). Metallothionein inhibits the erythrocyte hemolysis, stress induced ulcer and diabetes. In view of the high anti-oxidant activity, which is 7-8 times higher than human serum, it can be commercially utilized in sufficient quantity from seabuckthorn.

Chapter 4
Biochemical Changes During Fruit Ripening

Time for harvesting of seabuckthorn fruit varies with the variety and plantation region. Fruit of seabuckthorn should be harvested when it has attained the required ripeness (Antonelli *et al.*, 2006). Fruit of seabuckthorn is usually ripe and suitable for harvesting from mid August to the end of September depending on local growing conditions and type of cultivar. Fruit maturity in an orchard can also vary by up to 10-15 days. The fruit is ready to pick when bright yellow to reddish orange in colour. The fruit remain on the plant for a long time of the winter months. Harvesting time of seabuckthorn fruit also depends on the type of active compounds at peak. Various active compounds vary different with ripening of seabuckthorn fruit (Figures 1–3).

Organic acids and sugars constitute the main part of the soluble solid fraction of the soft parts of seabuckthorn berry. In general, there is a decline in soluble solids during matutaion of berries. Sugar components are important ingredients of berry, because they play an important role in determining the sweetness of the fruit, and in

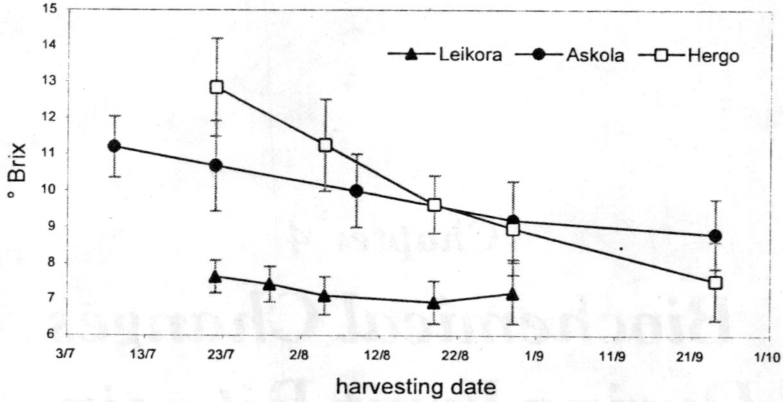

Figure 1: Soluble Solids Content Evolution Along Time in the Berries of Three German Xultivars (In: Antonelli *et al.*, 2006)

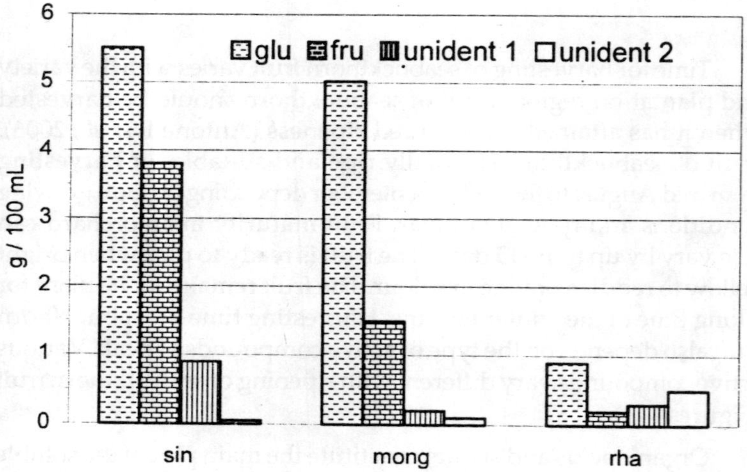

Figure 2: Soluble Sugars Content in the Raw Juice of Subspecies *sinensis*, *mongolica* and *rhamnoides*
***Source*: Kallio *et al.* (1999)**

tempering the tipical sourness of the juice; in fact the sugar:acid ratio has been reported to constitute the major promoter of taste of seabuckthorn berry juice. Vitamin C content declines rapidly after fruit has ripened.

While climatic conditions during growth and maturation may affect to some extent vitamin C accumulation, studies suggested decreasing trends in vitamin C concentration during fruit ripening. In a Finnish study (Kallio *et al.*, 2002)), a quite linear decline from the beginning of September to the end of November was observed in four different *rhamnoides* clones cultivated in Finland (Helsinki). In a study conducted on 6 clones from native stands in southwestern Finland, the content of vitamin C acid decreased in all clones by 25-60 per cent over a time span of 34 days from August 17 to September 21 (Rousi, and Aulin, 1977). However, carotenoid content has been found increasing with the maturation of fruit (Figure 3)

Total contents of tocopherols and tocotrienols in seeds showed a continuously increasing trend during the harvesting period (from the end of August to the end of November), which led to conclusion that the best time to maximize vitamin E is at the end of ripening time or even when the fruit starts to be overripe.

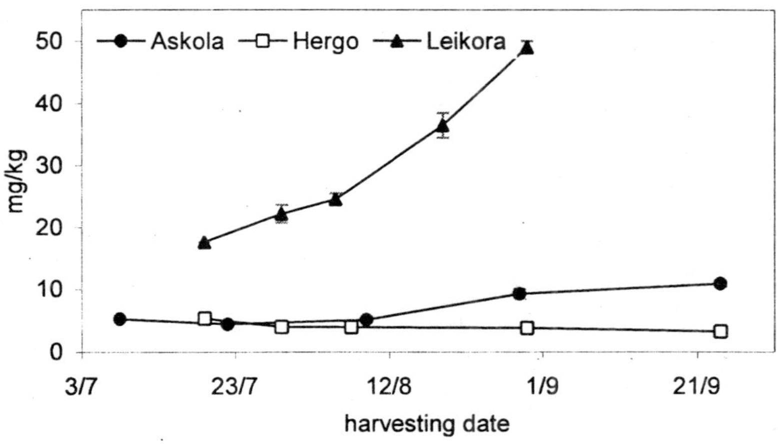

Figure 3: β-carotene Content (mg/kg Berry Fresh Weight) During Different Harvesting Dates (In: Antonelli *et al.*, 2006)

The principal sterol and sitosterol contents seem to decrease during the harvesting period. The main classes of phenolic compounds present in seabuckthorn berries are flavonols, proanthocyanidins, catechins (flavan-3-ols) and phenolic acids. Among the flavonoids, studies have found that the contents of quercetin and kaempherol in seabuckthorn fruits changed significantly during maturation, but in different ways. The overall mean of quercetin decreased 29 per cent in 19 days. In contrast to quercetin, the content of kaempherol increased significantly (33 per cent) in 5 cultivars. The oil content in whole berries is significantly influenced by the harvesting time, whereas, the annual variation has been reported to be extremely small. Application of inorganic fertilizers (N, P, K, in spring) increases oil content and decreases the content of carotenoids in the oil.

Chapter 5
Utilization in Health Industries

Seabuckthorn rich in anti-oxidants is being viewed as an important plant for preparation of a number of health protection juices, cosmetics and medicines. Seabuckthorn has been mentioned as a medicinal plant "Amlavetas" growing in Himalayas, as mentioned in ancient Indian System of Medicine, "*Ayurveda*" written during 500-5000 BC. The medicinal value of seabuckthorn berries was recorded in much detail as early as the 8th century in the Tibetan medical classic "*rGyud Bzi*" written during 'Tang Dynasti', which has 32 chapters on medical prescriptions for various health problems of blood circulation systems, skin wound, anti-inflammation and strengthen and coordinate the balance of functions among liver, stomach, spleen, kidney and heart. With scientific back up, seabuckthorn has become the source of important medicinal and nutritional products in Germany, Finland, Sweden and Asia and its nutritional and medicinal values is being noticed and explored in North America. The seabuckthorn industry has been thriving in Russia since the 1940s, when Russian scientists discovered its rich vitamin C value. Seabuckthorn fruit oil was also used for protection from radiation damage due to nuclear explosion in Chernobyl, Ukraine in 1986. Seabuckthorn juice was utilized in the diet of Russian cosmonauts. China started large-scale plantation and commercial utilization of seabuckthorn since 1984, producing a range health juices, cosmetics and medicines. More than 40 countries of Europe, Asia and North America are actively harnessing rich medicinal potential in health protection. India has also emerged as an important market for seabuckthorn products, a number of companies are heavily investing on seabuckthorn and producing a number of health juices and cosmetics. Many countries like USA, Korea, Japan and Bolvia have also introduced seabuckthorn for environmental conservation and commercial utilization. Globally, 8 International Conferences have been organized on seabuckthorn, since 1989.

Seabuckthorn fruit and leaves are rich source of anti-oxidants like vitamin C, E, carotenoids and flavonoids etc., which are known to neutralize the disease causing free radicals. Indian scientists discovered as early as 1962, the inhibitory effect of seabuckthorn on tumors, however, real credit to start the work on the modern medicines of seabuckthorn goes to Russian scientists during 1950s. Seabuckthorn oil was also formally listed in the Pharmacopoeia of China in 1977 and clinically tested in Russia, Finland and China etc. The oil and juice, commercially produced from its fruits and extracts from leaves, are used in Russian traditional and official medicine for the wound healing, antibacterial, as anti-ulcer and anti-inflammatory (antiphlogistic) agent and as the multivitamin product. The oil products from the seabuckthorn fruits and vegetative parts of plant possess hepatoprotective activity. In the official medicine of Russia, the plant extracts are commonly used as the components of various medicinal, compositions in dermatology, stomatology, ophthalmology, veterinary and cosmetology etc. A number of medicinal and cosmetic products, as well as food additives have been recently produced from the seabuckthorn extracts. It has found to possess anti-oxidant and immunomodulatory properties (Geetha *et al.*, 2002). Indian studies on seabuckthorn seed oil has found its strong efficacy in healing of wound, burns and gastric ulcer (Varshney *et al.*, 2005; Tyagi *et al.*, 2005). Indian studies have also proven its effectiveness in treatment of hypoxia, protection from radiation damage and treatment of cardiovascular diseases (Geetha *et al.*, 2002, 2007).

Seabuckthorn juice is a driving force of seabuckthorn based industries. Its fruit pulp contains up to 1 per cent oil and is rich in both lipophilic and hydrophilic antioxidants. The effect of seabuckthorn juice on plasma lipids, LDL oxidation, platelet aggregation, and plasma soluble adhesion protein were studied in healthy men taking seabuckthorn juice or placebo for 8 weeks. Seabuckthorn juice increased plasma HDL levels (by 20 per cent) and at the same time decreased the susceptibility of LDL to oxidation.

High content of organic acid (2-4 per cent) and vitamin C make it highly useful for beverage industries, particularly health protection juices. All the protein forming amino acids in free form are found in the juice, which make seabuckthorn food products highly nutritious and digestible. Seabuckthorn juices, being multi-vitamin, particularly in vitamin C, have become globally very popular for health protection of weak children, pregnant women, aged people and office goers.

Russian and Chinese scientists discovered four main pharmacological functions of seabuckthorn fruit oil, 1. Anti-bacteria 2. Diminishes inflammation 3. Anti-cancer and 4. Promotes regeneration of tissues. Several drugs have been made globally from seabuckthorn oil, curing a number of diseases like blood pressure, heart problems, cancer, ulcers, burns and aging etc. In India, presently, 20 processing industries have been established in seabuckthorn growing areas and a number of companies have been producing a range of juices, jams, jellies, tea, wines and number of cosmetics.

Based on its nutrients, many medical studies, testing its effects on the skin, showed that seabuckthorn extracts could effectively improve the micro-circulation of blood capillaries and nourish the skin and hair. Globally scientists utilized its nutrients and medical effects to develop several kinds of seabuckthorn beauty creams and cosmetics. Seabuckthorn oil is the raw mineral oil and natural nutrients with natural pigments. Seabuckthorn beauty creams are quite stable for a long period. Seabuckthorn beauty creams made the skin fair, clear and delicate. Studies have also shown that seabuckthorn fruit and leaf extracts can improve metabolism, retard skin maturation and made the skin smooth and soft. Many anti-aging and sun protection creams of seabuckthorn are now globally popular. Hair care also retarded baldness and promoted hair growth. Global trade in seabuckthorn products is believed to have crossed now over 2 billion USD. Following types of seabuckthorn products are being produced globally:

Seabuckthorn seed oil has been found to improve the blood microcirculation, helping in the skin structure maintenance. Seed oil has also been found to speed up the regeneration of damaged epidermal cells, keeping the skin smooth and young. As the oil has also been found to be anti-inflammatory, contributing in a general recovery of the damaged skin cells. These skin improving effects appear to be due to the presence of high contents of polyunsaturated fatty acids, carotenoids and sterols in seabuckthorn fruit oils, both pulp and seed oils. Seabuckthorn fruit oil can be used in a variety of cosmetic formulations, which are anti-aging, useful for improvement of microcirculation, has anti-oxidation protection and anti-inflammation and helps in epidermal regeneration and UV protection and therefore useful sun lotions. Now, a variety of beauticreams and cosmetics made from seabuckthorn fruit oil are in market globally.

Chapter 6

Fodder and Feed

Fruit residues, which account for about 25 per cent of the total fruit weight, are rich sources of protein, fat, amino acids, vitamin A, C, D, and E, carotenoids, flavonoids and micronutrients. Feeding on supplementary seabuckthorn foliage and fruit residues, pigs increased their health by 9-21 per cent, milk production in goats by 6-7 per cent, chickens enhanced their rate by 9-13 per cent and laying output by 25-29 per cent (In Lu Rongsen, 1992). In Zhungeer county, Inner Mangolia, one sheep or goat fed on seabuckthorn leaves produced 5 kgs more mutton or 30 per cent more of cashmere than that fed with other fodders. Live stocks in Lahaul-Spiti, H.P, India feed on its green foliage during the early winter, when all the fodder resources dry up. Indian study revealed that crude protein content of seabuckthorn leaves (18-25 per cent) is higher than red clover (17.1 per cent), and equal to white clover (24.7 per cent), whereas the fat content of seabuckthorn leaves (3-5 per cent) is significantly higher than both red clover (3.6 per cent) and white clover (2.7 per cent). The annual yield of fresh seabuckthorn leaves reach to 16 tones/ha, which is 2.5 times higher than alfalfa (Singh *et al.*, 1999).

Chapter 7
Fuelwood

Seabuckthorn is a promising green energy plant, because of its high quantity of biomass, high calorific value and quick regeneration after cutting. The plant meets the increasing need of firewood for the framer's daily life. A 4-year old plantation on marginal land produced 8-10.5 tones of firewood, equal to 6-8 tones of standard coal. Branches of seabuckthorn plants are important fuelwood in the whole cold deserts of India. In China, a 6-year old plantation of seabuckthorn produced 18 tons of dry fuelwood/ha/yr, which is equivalent to 12 tons of standard charcoal. The average calorific value of dry seabuckthorn wood is 4785 K calories/kg, which is more than most of the tree species. One ton of seabuckthorn wood is equal to 0.7 ton of standard coal (In Lu Rongsen, 1992).

Chapter 8
Environmental Conservation

Wide adaptation, nitrogen fixation, extensive and well-developed root system, compact canopy and dense forest of seabuckthorn, protect the soil from wind and water erosion. Seabuckthorn, because of its strong root system and nitrogen fixing ability is well adapted to both dry and soil conditions. Therefore, it is being used in reclaiming and conserving soil, especially on fragile slopes and riversides. There are several success stories of control of soil erosion by seabuckthorn forests in China. In Shanxi province, China, 74 km of seabuckthorn forest was planted on the bank of yellow river, which decreased sediment discharge by 3-5 million tons/yr (Min *et al.*, 1989). Seabuckthorn forest, when raised in a watershed area in Zhungar country, increased vegetation cover from 20 to 61 per cent, soil erosion decreased from 40,000 to 5000 tons/km^2/yr after 3 years of plantation in 1989. Grain yield increased from 160 to 373 kg/person and per capita income increased from 330 to 4260 rupees in the region. In some of the areas it reduced soil erosion up to 96.6 per cent. Local people called it 'Green hope' (In Lu Rongsen, 1992).

Seabuckthorn has an outstanding ability to take roots even in poor soils, because of its ability to fix atmospheric nitrogen through the presence of symbiotic bacteria *Frankia* (*Actinomycetes*) in the root nodules and add about 180 kg of nitrogen per ha per year, which is equal to soybean (Jike and Xiaoming, 1992). Seabuckthorn plants when planted in poplar stand, later achieved 4 times height and 3.7 times diameter than a pure stand of poplar. The nitrogen content of the cultivated layer (0-20 cm) and organic matter content in the mixed forest also increased by 11.5 and 26.6 per cent respectively, as compared to pure stand of poplar (Quanzhong *et al.*, 1989).

Seabuckthorn also plays an important role in the conservation of wild life. Many wild animals use seabuckthorn for food and shelter. In Losses Plateau region of China, 51 out of 350 bird species are entirely dependent upon seabuckthorn fruits as a food and 80 bird species are partly dependent upon seabuckthorn. A number of wild animals take shelter in seabuckthorn stands for protection (In Lu Rongsen, 1992).

Chapter 9
Climatic Requirements

Seabuckthorn naturally grows in Europe and cold regions of Asia. Therefore, plant is naturalized in snowfall or low temperature conditions, which meet the chilling requirement of the plant. Despite seabuckthorn's winter hardiness, it has high temperature requirement for active growth during summer. There is natural distribution of seabuckthorn in the regions with the temperature isolines over 25°C on the average in the summer. Therefore optimum temperature during summer for natural distribution of seabuckthorn is between 15-25 °C. The maximum radiation in the sunny days of May-July is 23.5-26.5 k cal/cm^2 between the isolines. This indicates, that seabuckthorn plants, though favour cold climatic conditions, however it requires a certain high temperature condition during the active growth period. Vegetative growth begins at average daily air temperature of 5-7 °C. It flowers at 10-15 °C and requires total effective temperatures, from spring to harvest time of September of 1445 °C to 1750 °C, depending on latitude/altitude and species of the seabuckthorn. Higher air temperature (25-30°C) during June-July months facilitates the accumulation of oil and sugar in fruit, whereas cool and rainy weather favours vitamin C formation. Latitude and altitude also increases carotene content, within limits, the higher the better. Seabuckthorn grows in low altitude regions of Europe and high altitude (2500-5300 m asl) in Himalayas and even it has been introduced at lower altitude (1500 m asl) in China. At the start of winter (November), plant defoliates for adaptation and go to dormancy conditions of winter, when frost hardiness, *i.e.*, resistance of the plant tissues and organs to low negative temperatures, is the highest in deep dormancy during November-January. During this time, negative temperature of -50 °C may be tolerated by seabuckthorn plant (In Lu Rongsen, 1992).

Chapter 10
Global Distribution

Seabuckthorn (*Hippophae rhamnoides* L.) is a plant of cold regions of Asia and Europe. It grows in high altitude areas in Asia (2000-5500 m asl), whereas, it grows in low altitude areas of Europe. It has been reported growing in about 40 countries like Russia, China, Indian Himalayas and Central Asia, Germany, Finland and Sweden. It has also been introduced in Canada, USA, Bolvia, Korea and Japan (Singh, 2003). Under natural conditions, seabuckthorn grows on hill slope, riversides, lakes, seashores and other wastelands, where conditions do not suit many other agricultural crops. Each species have been found to prefer certain specific site conditions. Although it prefers riversides due to sandy soil and adequate moisture, it is also well adapted to dry conditions, as it takes root easily even in dry and infertile soil due to presence of nitrogen fixing bacteria, *Frankia*, in its roots. Rich resources of seabuckthorn are found in China, Russia, Central Asia, Mangolia and India.

India

In India, seabuckthorn is found in Himalayan states of Himachal Pradesh, Ladakh (J&K), Uttranchal, Sikkim and Arunachal Pradesh. There are about 12,000 ha of seabuckthorn under thick forests, mainly in Ladakh. In Himachal Pradesh, seabuckthorn grows in Lahaul-Spiti, upper Kinnuar and Pangi in Chamba district. In Lahaul, it grows throughout the length of the Chandra valley in the villages, *i.e.* Sissu, Khorpani, Raling, Khongsar, Gondhla and Dalang. Several good stands of *H. rhamnoides* grow on both sides of Bhaga river form Gemur, Tinu, Jispa to Darcha. *H. salicifolis* is found in Kardang and Tinu villages. In Chandra-Bhaga (Chenab) valley, seabuckthorn is found as component of agroforestry systems in the villages Kirting, Sansa, Jhalma, Jhunda, Jasratha and Madhgaon. In Mayar valley, seabuckthorn is spread in Urgos and Chaling villages. The distribution of seabuckthorn is at 2500-3200 m asl in Lahaul.

Spiti, a cold desert valley of district Lahaul-Spiti, adjoins Tibet on its eastern border and Ladakh on its north. There are excellent stands of *H. rhamnoides* between Shego and Lara (3760 m), Lingthi (3889 m) and Shichling. *H. rhamnoides* have also been reported growing in upper areas of Spiti in villages like Kiato, Kiamal, Morang, Sumling and Rangrik. *H. tibetana* have been reported growing in higher altitude areas at Losar (4100 m) to Takcha (4500 m asl), Kibbar (4100 m) and Chicham (4200 m). Both *H. rhamnoides* and *H. salicifolia* have been found in Chango area of upper Kinnaur. Seabuckthorn has been also found in high altitude areas of Shimla and Kullu districts.

In Ladakh, a high altitude cold deserts, seabuckthorn (*H. rhamnoides*) has been reported growing in all the five valleys of Ladakh, *i.e.* Indus, Nubra, Zanskar, Changthang and Suru. It grows well on the riverbeds, riversides, and slopes. In Indus valley, seabuckthorn grows from Thiksey to Shey, Choglamsar, Spituk, Phyang, Nimmu, Bazgo and Saspol. Nubra valley is very rich in seabuckthorn, where it grows along the bank of Nubra and Shyok rivers in Warsi, Panamik, Sumur, Tegar, Bogdang, Turtuk and Pachathang areas. *H. tibetana* has been found growing in alpine zone of the region. Recent surveys have estimated over 11,000 ha under seabuckthorn in this region.

In Uttaranchal, mainly two species (*H. salicifolia* and *H. tibetana*) are reported, out of which *H. salicifolia* are more common and reported to exists in abundance in Garhwal Himalayas *viz.* Gangotri, Harsil, Sukhi, Gagnani, Hari ki Dun and Janki Chatti, Nand Chatti, Hanuman Chatti areas of Yamunotri in District Uttarkashi. In Rudraprayag district Gaurikund, Rambana and Kedarnath areas found to have large diversity. Pandukeshwar, Hanumanchatti, Ghagharia, Suraithota and Jhelum areas in district Chamoli found natural habituated regions of seabuckthorn. In Kumaon division seabuckthorn populations are reported from in Kali valley, Bogdiar, Gori valley, Budhi, Byanse and Darma etc. In north-east India, seabuckthorn grows in dry temperate regions in north district of Sikkim at an altitude of 2280-3100 m asl. Seabuckthorn has also been recorded in Dibang valley in Arunachal Pradesh, another state in northeast India.

China

About 1.2 million ha has been found under seabuckthorn in China, half of which has been artificially cultivated. Seabuckthorn resources are distributed in the provinces of Liaoning, Hebei, Shanxi, Shaanxi, Gansu, Qinghai, Sichuan and Yunan, inner Mongolia, Ningxia, Xijiang and Tibet. The altitudinal range varies from a minimum of 420 m above sea level in hilly region in western part of Liaoning to a maximum of 5200 m in Mount Qomolangma. Various regions in different provinces, which have natural seabuckthorn are: Zhaoyang region in Liaoning; Chende and Zhangjiako regions in Hebei; Luliang and Taihang Mountain areas in Shanxi; Huangnong and Laoshan mountain areas in Shaanxi; Qingyang, Pingliang and Tianshui regions in Gansu; Keshe, Hetian, Buerjing and Alitai areas in Xinjiang; Qilian, Datung, Gueide and Galmu counties in Qinghai; Ganzi and Arba regions in Sichuan and Zhongdian area in Yunan are the main sources of seabuckthorn. There are 6 species of seabuckthorn in China, *i.e. H. rhamnoides, H. salicifolia* D.Don, *H. tibetana, H. neurocarpa* and *H. gyantsensis* and *H. goniocarpa. H. rhamnoides* has four subspecies, namely ssp. *sinensis, yunnanensis, turkestanica* and *mongolica*.

Other Himalayan Countries

In Nepal, in this small Himalayan country, it was found that *H. salicifolia* is distributed between 2000-3850 m., whereas *H. tibetana* was found occurring between 2900-4500 m. Seabuckthorn forests were found in the higher areas, far away from the river system in Manang, Mustang, Dolpa, Humla, Mugu, Jumla and part of western Dolpa. *H. tibetana* occurs frequently in alpine tundra. In Bhutan, *H. salicifolia* was found in the trans-Himalayas at an altitude of 1500-3500 m asl. *H. tibetana* was found at altitude of 4000-5000 m asl. In Pakistan, seabuckthorn (*H. rham.*ssp. *turkestanica*) is distributed in 3000 ha in northern area of Pakistan. Its main forests lie in Baltistan, Gilgit (Ghizer and Hunza valleys etc.), Chitral (Mastuj, Garam Chasma etc.) and Swat (Gabral) at an altitudinal range of 2200-2800 m asl. There are 3000 ha natural forest of seabuckthorn in Pakistan. Having conditions similar to that of Pakistan, *H. rham.* ssp. *turkestanica* is distributed widely at a range of 2300-3000 m in mountains of Bamian, Kabul, Parwan, Gulbahar and Nuristan in Afghanistan.

Russia

Seabuckthorn is widely spread with a lot of morphological variations over vast territory of Russia, i.e Kaliningrad region, subcaucasus, southern part of Siberia (Altai) and eastern Siberia. There is a large area under seabuckthorn in Russia, particularly in Siberia. Siberia comprises 57 per cent of the total area of Russia. The region has many useful populations of seabuckthorn especially in Altai area. At present, seabuckthorn (ssp. *mongolica*) is cultivated as a fruit crop from Arkhangelsk in the north, though Tomsk and Chita to southern part of the country.

Central Asia

In the countries of former Soviet Union, seabuckthorn grows in Kazakhstan, Caucasus, Uzbekistan and Azerbaijan. Several surveys were carried out on seabuckthorn resources in the valleys of more than 70 rivers, coasts of the Issyk-Kul, Sevan lakes, the Capsian and Baltic Sea Shores. This whole region has vast and different types of seabuckthorn resources. In Azerbaiijan, seabuckthorn grows naturally in valleys of rivers of Kishchay, Girdimanchay, Lenkoranchay and Sara peninsula. Vast resources of seabuckthorn have been found in Kazakhstan, where it was found growing in Altay, Zaisan hollow, Djungar Alatau, north Tianshian, Zailyisky Alatau, Cungey Alatau, Ketmen range and Kara-Tau range, west Tian Shian, Kirgiz range and Issik-Kul lake valley.

Natural seabuckthorn brushwood areas in Mongolia are large and it covers more than 20,000 hectares. Western part of Mongolia has the main natural area of seabuckthorn. In the northern part of Mongolia, it is isolated and spread only on the banks of the rivers, Selenge and Orhon. In the western areas, seabuckthorn is found at the steppe (Bogd-Borh river), desert-steppe (Zavhan, Hovd, Tes rivers) and desert (Bulgan river) zones. In the northern areas, seabuckthorn is found at the altitude of 600 m, and in the western areas, 1400-2000 m above see level. There are two different subspecies of *Hippophae* in Mongolia, *H. rhamnoides* ssp. *turkestanica* and *H. rhamnoides* ssp. *mongolica*. Populations of seabuckthorn growing in the regions of Bulgan and Hovd river belongs to the *H. rhamnoides* ssp. *turkestanica* and populations of the other areas belongs to *H. rhamnoides* ssp. *mongolica*.

Europe

Except Russia, European countries lack the adequate resources of seabuckthorn, as most of the resources have been destroyed due to developmental activities, however plant is widely scattered in Europe.

Seabuckthorn widely grows in Germany. *Hippophae rhamnoides* ssp. *rhamnoides* (syn. *H. rhamnoides* ssp. *maritima*) occurs along the coast sides of north sea and Baltic sea. Homogeneous populations can be found on the islands, Hiddensee and Ruegen within the Baltic sea. *Hippophae rhamnoides* ssp. *fluviatilis* is distributed mainly outside of Germany, from Suisse to Italy and France. Only the northern part of its population can be found in South Germany (near river Rhine and in the Bodensee area). The third subspecies *Hippophae rhamnoides* ssp. *carpatica* is distributed between the rivers Danube and Isar, also in the southern part of Germany. Natural distribution of seabuckthorn has been reported in Finland, from Aland Inslands up the west coast along the Bothanian gulf to the valley of the Tornio river in the north. However, as early as 1960s, seabuckthorn strains from various European and Asian countries, were introduced at the Institute of Horticulture, Piikkio. Seabuckthorn of this area belongs to *H. rham.* ssp. *rhamnoides*. In Ukrain, seabuckthorn has been reported in delta of Danube, Ukrain, which has some 250 ha area under the species.

It is common in south and rarely found in the north Sweden. It is distributed on the Swedish coast along the Gulf of Bothnia and west coast of Sweden. It is found on the Baltic coast of Poland. However, it is being cultivated for reclamation of wastelands and industrial dumps. Hungry has introduced a number of improved cultivars, which are being cultivated at small scale. It has also been planted for the reclamation of the red mud, which contains industrial wastes of oxides of metals. While in the past, seabuckthorn was quite common in Norway, it has been found in few sites on river banks and along the coasts from Trondheim in the south to Harstad in the north. In southern part, it also grows on one site Boverdalen at an altitude of 1100 m.

In Czech, information about the natural distribution is not available in this country, but its orchards have been established near Velky Osek about 50 km to the east of Prague. In Netherland, seabuckthorn (*H. rham.* ssp. *rhamnoides*) has been reported growing as a pioneering shrub species on calcareous coastal sand dunes like places Meijendel, near Hague. Seabuckthorn has been found growing in eastern and southern sub-Carpathians in Danube delta area and black coast on the sunny river banks in Romania. Surveys have been carried out in south-eastern Romanian Carpathian mountains and several biotypes were identified. In Britain, seabuckthorn distribution has been reported on the sea shores around Britain. Naturally, it was found growing on the east coast of Britain, from Berwickshire to Kent.

North and South America

Seabuckthorn was first introduced at Dropmore, Manitoba, Canada by Dr. L. Skinner in 1930s from Siberia, Russia. The exotic form was identified to be *H. rham.* ssp. *mongolica*. In most of the sites, it has been raised as an ornamental plant, whereas it has been planted in Prairies for the conservation of wild life, reclamation of wastelands and control of soil erosion. For this purpose, PFRA Shelterbelt Centre at Indian head in Saskatchewan has distributed over one millions plants of seabuckthorn. In 1999, more than one hundred thousand seabuckthorn (120 kilometers of tree row) were planted in farmstead shelterbelts. With the realization of rich medicinal properties of seabuckthorn fruit and leaves, seabuckthorn has been introduced in some parts of United States of America. Seabuckthorn has been introduced in dry high lands of Bolvia, a south American country. Seedlings introduced from China although showed slow growth, but survived, whereas they showed fast growth at Patacamaya Research Station and had fruits, which indicate that seabuckthorn can be cultivated under the conditions of this country.

Chapter 11
Taxonomy

Seabuckthorn (*Hippophae* L.) is one of the 3 genera in the small nitrogen fixing family *Elaeagnaceae*. The taxonomy of seabuckthorn is based on the work of Rousi (1971), published as *"The genus Hippophae-A taxonomy study"*, supplemented by other taxonomists from China (In: Yongshan *et al.*, 2003).

1. *H. salicifolia*

Its trees are medium to tall in height (4-7 m). It has a willow like appearance. Leaf's margin revolute, adaxial surface densely covered with stellate hairs. Mid rib is clear. It is mild thorny. It is also found in Xizang (Cuona, Yadong, Jilong), Pakistan and Nepal. In India, is distributed in Lahaul, Kinnaur and upper areas of Uttrakhand in Himalayas. The altitude of distribution is 1500-3200 m.

2. *H. rhamnoides*

2a. Subsp. *sinensis*

It is a small tree or shrub, 2-18 m high. It is found in hills, on mountain slopes, valley bottoms, river banks or river beds of dried up, between 400-3100 (3700) m in China. It is very common in the Loess Plateau. It is the main resource for seabuckthorn development in China. It is found in Sichuan, Qinghai, Gansu, Shaanxi, Ningxia, Inner Mongolia, Shanxi, Hebei, Beijing and Liaoning. The current year branchlets are stiff. Leaves are opposite or subopposite. There are scale-hairs on the lower surface, white or rarely rusty-red. Most of fruits are tangerine. Seed surface in this type is shining.

2b. Subsp. ***yunnanensis***

This subspecies is found in valley bottoms, river banks and woodlands in foothills at an altitude of 2200-3700 m in China (Yunnan, Xizang, Sichuan, Qinghai (Nangqian alt. 3700 m). It is a tree or small tree, upto 20 m high. The current year branchlets are soft. Its most leaves are alternate, the midrib on the above impressed and up to top appearing as deeper sulcus, with most rusty-red scale-hairs on the lower surface. Its fruits are yellow; sometimes the carpodermis is hard to be separated from seed coat. Its seed are usually flat.

2c. Subsp. ***turkestanica***

It is a hardy shrub or small tree 3-7 (15) m in height. It is found in India (Himachal Pradesh, Ladakh), China (Xizang, Xinjiang, Gansu), Afganistan, Tadzhikistan, Kirghizia, Kazakhstan and Uzbekistan. Epidermis of branches is silvery, with more and usually branched thorns. Leaves are alternate, narrower, 2-4 (5) mm wide, both surface silvery. The length of most fruits greater than breadth. Fruits are reddish-orange in colour.

2d. Subsp. ***mongolica***

It is a 2-6 m high shrub. It is found in river terrace or flood land, at 1200-1800 m in China (Altai mountains), Mongolia, Russia and Eastern Kazakhstan. Branches are brown, with less and no branched thorns. Leaves are alternate, the widest usually above the middle, 5-8 mm wide, above green, adaxial surface silvery, acuminate at apices. It is commercially cultivated in Russia.

2e. Subsp. ***caucasia***

It is a tree, usually up to 10 m high. It grows along river banks and in flood lands, between 1000- 2500 (3000) m in Azerbaijan, Georgia, Armenia, Russia, Turkey and Iran. Branches are brown, with less and no branched thorns. Leaves are alternate, the widest commonly below the middle, 3-7 mm wide, abaxial usually green, adaxial surface silvery, acuminate at apices.

2f. Subsp. *carpatica*

It is a shrub or small tree. It is found in valleys of Alps, Carpathian mountains, Danube and the shore of the Black Sea, from sea level to 380 m. It is found in Romania, Ukraine, Yugoslavia, Hungary, Austria and Germany. Branches are straight. Leaves are alternate, 5-7 mm wide, adaxial surface is usually mixed with most rusty-red scales. Fruits are 6-8 mm long, 5-7 mm wide.

2g. Subsp. *rhamnoides*

It is a shrub or small tree. It is found growing on sea shores and valley bottoms, from sea level to 1100 m in Poland, Germany, Denmark, Sweden, Finland, Russia, Norway, Netherlands, Belgium, France and Britain. Its branches are more or less curved. Leaves are alternate, mostly 5-8 mm wide. Fruits are 8-11 mm long, 5-7 mm wide. Seeds are compressed.

2h. Subsp. *fluviatilis*

It is a shrub or small tree. It grows in open land, river bottoms, river banks and hillsides, between 100 1900 m in Austria, Italy, Switzerland, France and Spain. Leaves are alternate,3-6 mm wide, adaxial surface is usually mixed with most rusty-red scale-hairs. Fruits are 5-6 mm long, 4-6 mm wide. Its seeds are ovate.

3. *H. goniocarpa*

3a. Ssubsp. *litangensis*

It grows in valley terraces, hills, at 3700 m altitude in China (Sichuan, Litang).This subspecies differs from the typical subspecies by the young branchlets and the lower surface of leaves densely covered with stellate hairs, the leaf margin usually manifestly revolute, the midrib on the above side impressed and up to top appearing as deeper sulcus; the fruits tangerine or dark tangerine, 6.0-7.6 mm long, 4.5-5.3 mm wide, 1.4 times longer than wide.

3b. Subsp. *goniocarpa*

It is a small tree, 5-8 m high. It grows on mountain slopes, river banks, flood lands and valley terrace, at elevations of 2650-3650 m in Sichuan (Sunpan, Ruoagai, Hongyuan, alt. 3500-3650 m), Qinghai (Qilian, alt. 2500-3200). Branches are soft, branchlets of the current year red-brown or dark brown. Leaves adaxial surface are densely covered with scale-hairs, leaf margin usually explanate, never manifestly revolute, the midrib on the above impressed into sulcus but up to top shallow or absent. Buds of male flower are cruciform, of female flower near cruciform (bifid ovate, the second pair of the bud scales is clearly visible). Fruits are *terete*, juicy, apricot-colored or straw-yellow, 6-10 mm long, 4-5.9 mm wide, 1.45-2.1 times longer than width. Seeds are slightly flat, with 3-5 inconspicuous longitudinal angles.

4. *H. gyantsensis*

It is a small tree 5-8 m high. It grows in river banks, flood land and valley terrace, between 2600-5000 m in China (Xizang) and Sikkim in India. Its branches are soft, branchlets of the current year brown-yellow. Leaves abaxial surface are scattered stellate hairs and scale-hairs, adaxial surface is densely covered scale-hairs. Flower buds are ovate bifid. Fruits are yellow, longitudinal angles almost developed in towing-shaped.

5. *H. neurocarpa*

5 a. Subsp. *stellatopilosa*

It grows on river banks, flood land and river terrace, between 3400-4400 m in Sichuan (Daocheng, Litang, alt. 3700-4000 m); Xizang (Jomda,river banks of Yuqu Riverin Basu, Zuogong, Mangkang, Leiwuqi, Lhasa, alt. 3400-4400 m); Qinghai (Nangqian, Yushu) in China. This subspecies differs from the typical subspecies by the young branchlets, and the adaxial surface of leaves densely covered with stellate hairs. The leaf margin usually manifestly revolutes, the midrib on the above impressed and up to top appearing as deeper sulcus; the fruits faintly orange-coloured or yellow-brown, 5.6-6.5 mm long, 2.5-3.1 mm wide, 2.1 times longer than wide.

5b. Subsp. *neurocarpa*

It grows on river banks, flood land and river terrace, between 2700-3900 m in China (Sichuan, Qinghai, Gansu). Plants are 1-3.5 m high, the crown cover appearing is as platform in the adult plants. Branches are stiff, Branchlets of the current year are gray-white. Leaves adaxial surface are densely covered with scale-hairs, commonly explenata never manifestly revolute at margin, the midrib on the above impressed into sulcus but up to top shallowed or vanished. Flower buds are ovate or ovate bifid. Fruits are black-brown, curved prism, with little or very little juice, one end thinner, with 5-7 longitudinal angles, 7.8-8.4 mm long, 2.8-3.3 mm wide, 2.5 times longer than wide.

6. *H. tibetana*

Plants are dwarf, 7-60 (80) cm high. It grows on river bottoms, on river banks and steppes on higher mountains, at the elevations of 2700-5300 m in China (Sichuan, Xizang, Qinghai, Gansu), Nepal, India (Himachal Pradesh, Ladakh, Sikkim). Branches are pointing upwards, usually broom-like. Leaves are whorled, linear. Flower buds are ovate or ovate bifid. Fruits are dark tangerine, with 5-9 brown-black stellate ornamentation at apices, fruits and seeds.

In India, four species have been found, *i.e. H. rhamnoides* subspecies *turkestanica* Rousi, *H. gyantsensis*, *H. salicifolia* D.Don, and *H. tibetana* schelechtend. In Himachal Pradesh, the three species (*H. rhamnoides* subspecies *turkestanica*, *H. salicifolia* and *H. tibetana*) are found, whereas in Ladakh, *H. rhamnoides* and *H.tibetana* are found. In Uttrakhand, *H. salicifolia* and *H. tibetana* are reported. *H. gyantsensis* is found in Sikkim.

Chapter 12
Breeding Strategies

Wild populations of seabuckthorn are generally characterized by undesirable characteristics like occurrence of thorns, low fruit yield, low oil content and infestation of diseases and pests etc. Growers need to collaborate with seabuckthorn breeders for the development of promising forms. Large variations in morphological and biochemical characters in seabuckthorn natural populations provide us an opportunity for the genetic improvement of this plant. Work on breeding of seabuckthorn has been carried for long in former Soviet Union. During the last one-decade, researchers have also been working on genetic improvement of seabuckthorn in China, Sweden, Germany, Finland, Canada and India.

Work in Russia is a classical example to study the breeding on seabuckthorn. In Russia, there has been special emphasis on the selection of female forms possessing good characteristics like absence of thorns, dwarfness, compact crown, winter hardiness, large fruit (>0.3 g), high fruit yield (>5 kg/plant), strong fruit skin, solid pulp and long peduncle (>3 mm), easily detachable from the branch, high oil content (>5 per cent), high vitamin C (>100 mg/100g) and carotene. The male forms were selected for winter hardiness and their flowering synchronize with flowers of female plants. After evaluation, Russian workers have made selections as first improved forms by the names of *Novost, Altaya, Dar Katuni, Zolotoi pochatok, Maslichnaya, Vitaminnaya* and *Chuyskaya* (Kalinina, 1988).

The whole breeding programme of seabuckthorn in Russia is based mostly on single subspecies *mongolica*. Since 1933, more than 23 improved varieties have been developed at Lisavenko Institute of Horticulture for Siberia at Barnaul, Novogorod Institute, formerly Gorki Institute and some other institutes. Fruit yield varies from 7.5 to 18.0 t/ha, weight of 100 fruits varies from 62 to 120 g (maximum of 140g), length of peducnle 3-6 mm, oil content 4-8 per cent, carotenoids 15-48 mg/100g, sugar 5-10 per cent and acidity 1.0-1.9 per cent. The most of the varieties are thornless or mild thorny. Three methods were applied in breeding of seabuckthorn, *i.e.* selection, hybridization and mutagenesis.

Other research organizations engaged in breeding of seabuckthorn are The Berry and Experimental Stations of the Siberian Branch of the Lenin All-Uninon Academy of Agricultural Sciences in Buryatia and Novosibirsk, Institute of Cytology and Genetics, Novosibirsk, which have produced 5 varieties through selections since 1985. Botanical Garden of Moscow State University and other research organizations in former Soviet Union have developed 14 varieties. Durińg late 1980s, breeders produced 37 varieties, among them 9 have been introduced for cultivation in different regions of former Soviet Union and other countries like Hungry, Poland, Bulgaria, Yugoslovakia and Mongolia. But forms from Altai are susceptible to withering (due to *Fusarium* wilt) and seabuckthorn fly. A planting spacing of 4 m x 2 m has been recommended for Siberian seabuckthorn cultivars (Kalinina, 1988).

Some forms like Altai cultivars (*H. rhamnoides* subsp. *mongolica*) in Russia, are unable to withstand long thawing weather, thus prone to cold damages, whereas *H. rhamnoides* subsp. *rhamnoides* is known to have more stability under long thawing weather. Therefore, *H. rhamnoides* subsp. *mongolica* has been recommended for cultivation under moderate climatic conditions of Siberia, the Far East, the European part of Russia, the north of Mongolia, northeast China etc. unlike in southern regions of the country where vegetation begins quite early and strong frosts are marked during flowering of seabuckthorn.

Russian experts recommend for planting 2 or 3 cultivars together, differing in ripening period by 10-15 days. There is yet to produce any late ripening variety for fruit collection in winter. There may be significant losses of vitamins etc. due to periodic frosts and thawing of fruits. Attack of the crop by birds can also reduce the fruit yield by 30-50 per cent. Fruit yield may be high but due to lacking of application of modern management practices, productivity of Russian plantations is about 10 tons fruit per hectare, with an annual profit of $ US 3000 per hectare.

In China, improvement of seabuckthorn started with surveys of natural populations in large parts of country and selections of promising forms of seabuckthorn. Six *ex situ* gene conservation plantations of provenance and plus tree were established in Dengkou, Fuxin, Yongshou, Xinin, Tianshui and Xifeng. A set of superior exotic Russian cultivars of provenances was introduced to various ecological regions, largely enriching genetic diversity of the species in China. 14 superior cultivars of provenances were selected at different levels, from natural species or introduced exotic populations derived from seedlings. Some of these cultivars with high economic values, "Wu-lan-sha-lin", "Liao -Fu-1', "Liao-fu-2" etc. are thorn less types, with large fruits, high contents of oil and high fruit yields, have been used in operational mass production. "Jifeng" and "Jida" are high yielding cultivars with a fruits yield of more than 20 kg per tree (Quan, 2003). China also introduced a number of high yielding Russian cultivars, some of them were found suitable for Chinese conditions.

Then hybridization between improved Russian cultivars (*Hippophae rhamnoides* subsp. *mongolica*) and Chinese selections (*Hippophae rhamnoides* subsp. *sinensis*) have been carried out. Russian forms were early flowering, high yielding with large fruits and mild thorny, where as local forms were widely adapted and thorny. The F1 hybrid showed very promising results in F1 population (Zhengpin *et al.*, 2007). Similar trials are in progress in many countries like Germany, Finland and Canada.

The pioneer works on the experimental mutagenesis in seabuckthorn were initiated by Privalov in 1959, at the Institute of Cytology and Genetics SD AN of the USSR. Seeds, collected from wild maternal plants of Altai population (the Chulishman river), were treated with gamma-irradiation at a dose of 15 kR (Privalov *et al.*, 1971). Promsing results were not produced initially. Seeds, were exposed to subsequent treatment with the chemical mutagen nitrosomethylurea (NMU, 0.01 per cent). In 1992, 'Ziryanka' was released as a commercial variety. The majority of the following varieties were created by the Institute of Cytology and Genetics of SD RAS in collaboration with Novosibirsk Zonal Fruit and Berry Growing Experimental Farm: Druzhina, Podruga, Zolotoy kaskad, Ivushka, Ognistaya and Zarnitsa.

Presently, *H. rhamnoides* ssp. *mongolica, rhamnoides, sinensis* and *tukestanica* and *H. salicifolia* have emerged as most promising for raising of orchards. *H. saliicfolia* and ssp. *sinensis* are most rich in vitamin C, whereas, ssp. *mongolica* is high yielding and rich in oil, whereas ssp. *turkestanica* is hardy, widely adapted, rich in oil, vitamin E and beta-carotene. Based on the Russian and Chinese experiences on seabuckthorn, following are some important characters, which are required to be considered for making selections from natural populations or cross hybridization between promising forms of seabuckthorn:

Thorns

In highly thorny forms, number of thorns may vary from 3 to 7 thorns per 10 cm of 2 or 3 year old branches of the plant. Russians have made natural selections for forms with lesser number of thorns. Through breeding, they could also obtain 5 thornless varities, *i.e. Novost Altaya, Vitaminaya, Velikan, Prevoskhodnaya* and *Alei* and 13 mild thorny varieties.

Yield

Fruits are produced on the branches, borne a year earlier. The fruit yield depends on the size of fruits, their density on the branches, number and length of fruit bearing branches. Lisavenko Institute has the distinction of producing all high yielding varieties of seabuckthorn. For example, *Novost Altaya* and *Dar Katuni* varieties produced 11.3 tons and 12.7 tons per ha, respectively, for over 10 years of period. The institute has recommended the following high yielding female improved varieties for orchard raising: *Vitaminnaya, Dar Katuni, Shcherbinki-1, Chuskaya, Oranzhevaya* and *Obilnaya* and following male varieties: *Katun* and *Sayan*. However, there are high yielding varieties which produce fruit as high as 20-49 tons/ha in Russia and China.

Fruit Weight

Breeders in order to increase yield, looked for large fruit size, especially large fruit pulp size. As a result of breeding work, they could produce the fruits with fruit weight of 100 fruits up to 60-110g.

Fruit Skin

Fruit skin toughness is an important trait, which affects the suitability for mechanical harvesting of seabuckthorn and its transport. Skin damage during fruit harvesting invites the growth of microbes, therefore, affecting the quality of raw materials and also causing crop losses.

Fruit Quality

Quality of fruit is an important parameter in seabuckthorn breeding programme. Seabuckthorn fruit is well known to be quite rich in vitamin C. Other compounds of interests for a breeder may be vitamin E, flavonoids, carotenoids and unsaturated fatty acids. Russian breeders could produced improved forms with high oil content of 4.7-6.9 per cent, 50-330 mg/100g vitamin C and 2.8-7.6 mg/100g carotene. Vitamin C rich varieties are *Oranzhevaya, Sibirskaya* and *Yantarnaya, Zolotistaya Sibiri*. Carotene rich forms are Luchezarnaya, Panteleevskaya, *Yantarnaya* and *Zolotistaya Sibiri*.

Fruit Peduncle

Pedicle length and abscission affects the efficiency of mechanical harvesting. Therefore, fruits with long peduncle (>4 mm) with easy separation are desirable for selection. Forms were selected with not only larger fruit size, but also easily detachable by mechanical means. On crossing the Chuyskaya, Kuddyrga-1 and other varieties with Katun ecotype, Russians could produce promising hybrids. Improvement of length of peduncle also helped in easy separation of fruits.

Harvesting Period

In order to expand harvesting period, breeders have produced early and late harvesting varieties of seabuckthorn. Consequently, they could extend the harvesting to 2.5-3.0 months of fruiting period, *i.e.* from second half of July to the middle of October. Further, varieties are also being developed, which can be mature and develop fruits during winter, while maintain their biochemical quality.

Pest and Diseases Resistant

For the higher yield and better acceptability in market, breeders have given special attention to development of varieties, which are resistant to pests and diseases (*e.g. Fusarium* wilt).

Winter Hardiness

Seabuckthorn, which belong to temperate zone, may or may not tolerate winter frost. In some regions of Siberia, floral buds of male plants of seabuckthorn are often frostbitten, sharply decreasing the number and viability of pollen, causing reduction of fruit yield. After evaluation of floral buds of 10 hybrid forms of male seabuckthorn, Russian breeders could select *Katun* and *Alei* varieties.

Dissert Varieties

Seabuckthorn has demand in fresh juice, juices with out addition of sugar, beverages, jams etc., therefore, considering requirement of health food products, dissert varieties are being developed under the breeding programme in Russia.

Taste

Taste of seabuckthorn fruits is an important parameter for utilization in various products development. However, the sugar-acid index (SAI) can be used to decide the taste of seabuckthorn fruits. On the basis of the studies, the experts have made a table to decide the taste of the fruit on the basis of sugar-acid index.

Pollen

The number of pollens and duration of their release, compatible with the development of female flower, directly affect the fruit yield. Therefore, it is an important parameter for selections that male plants produce optimum quantity of pollen at the time of female plants are in flowering stage with receptive stigma.

Chapter 13
Propagation

Seabuckthorn can be propagated from seeds, hardwood and soft wood cuttings, layering and suckers.

PROPAGATION FROM SEEDS

This method is generally used to propagate seabuckthorn for meeting large-scale demand of its plants for the environmental conservation and control of soil etc., where no control of male and female plants is required. The method is comparatively easy as compared to other methods of propagation. Seedlings produced from seeds are both male and female and it takes about 5 years in fruiting and identification of male and female plants of seabuckthorn; therefore this method is not used for raising commercial orchards.

Seeds, for raising vigorous nurseries should be collected from healthy plants with desirable characteristics. Seed comprises 5-15 per cent of the total weight of fruit. 100 kg of seabuckthorn fruits may provide 5-15 kg of seeds. Seeds of seabuckthorn are minute (2-3 mm in length). There are 40,000-1,25,000 seeds in one kilogram of seeds, depending on size of the seeds. This method produces a greater number of plants and involves lower costs than other methods. If the germination rate reaches to 50 per cent, one kilogram of seeds can produce 20,000-60,000 seedlings of seabuckthorn.

Time of Collection

Seeds are separated in the stage of full biological maturity. For seeds collection, fruits are collected from the plants during late August to October. At the beginning of the browning phase of seeds, the majority of seeds are capable of germination. It is better to collect overripe fruits during September-October for higher seed germination (Eliseev and Mishulina, 1972).

Separation of Seeds

Fruits are harvested from the plants by machines, small tools or hitting by a stick. Seeds are separated by maceration and floating off the pulp and juice. Fruits can be collected in polythene bags, containing several pores at the bottom and its sides. After the extraction of juice, crushed fruits are soon sun dried for 3-5 days on a sheet. Mixy, generally used in home kitchen, easily separates seeds from the sun dried fruits. For large-scale separation, a commercial mechanical finishing machine may be used. Russian workers developed a machine for seed separation from the fruits (Poluparnev and Dulgov, 1988). Weak, damaged and diseased seeds are discarded. Then seeds are dried in the sun for a day.

Storage

Well clean and dried seeds (6 per cent moisture) are stored in sealed poly bags under dry and dark conditions or containers at low temperature. Generally, at room temperature, viability of seeds is quite high (85-100 per cent) for 2 years (Fefelov and Eliseev, 1986), but then it declines rapidly. The storage at low temperature for long period does not affect the viability, however, storage under high humidity badly affects their germination.

Pre-treatment

Germination of seeds under field conditions is influenced by various factors like source and maturity of seeds, dormancy, soil media, sowing depth and soil environment. Seabuckthorn seeds have short term physiological dormancy, therefore they take long time in germination under field conditions. To enhance early sprouting, seeds may be given some treatments. Pre-seeding treatment has been reported to break seed dormancy and greatly improve the seed germination. To solve the fungal problem, seeds are surface-sterilized in a solution of bleach (1:9:: bleach: water) for 5 minutes.

Chilling Treatment

Internal seed dormancy can be broken by placing the seeds at 5°C for 3 months (Siabough, 1974). For stratification, the seeds are mixed with wet coarse sand in the ratio of sand:seeds:3:1 and placed in a refrigerator at 0-5° C for 30-40 days. Seeds are mixed twice a week.

Water Treatment

To improve the seed germination, seeds are put into the water and floating seeds are removed. A high rate of germination under field conditions can be obtained either by soaking the seeds in water at room temperature for about 7 days, unless the seeds are semi-germinated (Lu Rongsen, 1992). This technique is useful when sowing of seeds are carried out in spring or indoors in a greenhouse.

Chemical Treatment

Soaking of seeds in KNO_3 (0.01 M) at room temperature (20°C) for 2 days greatly improves the seed germination. Some studies have reported better germination, when the seeds were treated with 800 ppm GA3 for 24 hours or 1N H_2O_4 for 2 minutes (Avanzato *et al.*, 1987).

Moist Sand Treatment

Stratification by soaking seeds in moist sand for 3 months at 5°C may also break internal dormancy (Siabaugh, 1974).

Bleach Solution

In order to prevent fungal infection of cotyledon after germination of seeds, a 20 minute soaking in 10 per cent bleach solution before sowing in the nursery beds is useful. To prepare the 10 per cent bleach solution, we add 1 part bleach in 10 parts of water. Beach is a 5.25 per cent sodium hypochlorite solution, which is the normal concentration in household bleach.

Drying

After soaking of seeds in water or chemical solution etc., seeds are air dried in sun for a day before sowing.

Nursery Beds

Seabuckthorn nursery should be raised in a place close to the expected plantation site. There should be enough sunshine, irrigation and transportation facilities. Before the seedbeds are prepared, better before onset of winter, leveling of the site is done, sufficient banyard / FYM manure is mixed in the soil and deep ploughing is done. Generally, the seedbed is 3-10 m long and 1 m wide, which is convenient for irrigation, weeding and other operations (In Lu Rongsen, 1992). For a better growth of seedlings, a soil mixture is prepared in the nursery with an organic content of 3-4 per cent and pH of 7-8.

Time of Sowing

Seeds can be sown both before and after winter. In Russia, workers found result of autumn sowing better in western Siberia. Both in Moscow and western Siberian regions, early vernal sowing of stratified seeds had sprouts in 7-10 days, whereas dry seeds germinate in one month. In Canada, the non-stratified seeds are sown in late September at a depth of 1 cm and a rate of 100 seeds / m in rows 60 cm apart at the PFRA shelterbelt Centre, Saskatchewan, Canada (Li and Schroeder, 1996). In the following spring they obtained a high rate of 90 per cent germination. Following the melting of snow in spring (April), soil temperature (15-20°C) at the depth of 5-10 cm is suitable for sowing of seabuckthorn seeds. Chinese experts recommend the seeds showing in spring (April), when the temperature of the soil at 5-10 cm depth, rises to 15-20° C (Lu Rongsen, 1992).

Sowing

Effect of sowing depth is important on seed germination. Surface sowing of seeds have been reported to have better germination than sowing at 1 or 2 cm soil depth (Li and Wardle, 1999). Seeds should not be sown deeper than 1-2 cm. When the sowing is carried out in late spring, the seeds should be covered by a thin layer of soil. Soil should be regularly irrigated with a spray or mist for avoiding mortality and better growth of seedlings.

Lebeda (Personal comm.) recommend sowing of seeds in a fertile site at a depth of 1-2 cm in late autumn and 5 cm in early spring. Sowing of seeds by tractor is also carried out, which may use 3.5 g per 1 m of furrow. Seeds are covered by thin (1.0-1.5 cm) layer of mixture of humus: sand::1:1, while distance between furrows is maintained at 18-20 cm, between seeds at 1.5-2.0 cm. He recommends dry seeds for late autumn sowing to avoid infection of fungi. During the spring sowing he recommend wet seeds. Ditch and drill seeding is generally practiced in northern China. The drill furrow of 4 cm width and 2.5 cm depth are made in the nursery beds. The drill spacing is maintained at 20-25 cm. The seeds are sown in the drills and covered with 1-2 cm of soil layer. Studies have found that pre-treated seeds should sown at the depth of 1 cm at a rate of 150 seeds/m^2 in during April or May; whereas non-treated seeds are sown in late September or early October.

At Gaozhung village, Gucheng town, Pengyang, China, nursery sloppy land was ploughed, removing the weeds. 30-37.5 tons of manual fertilizer, 112.5 kg of urea, 750 kg of phosphate fertilizer were applied per hectare. In mid July, seeds were treated with hot water (100°C) for 2 minutes and then added cool water for 2 hors of soaking. After air drying, seeds were sowed at a rate of 180-225 kg per acre with drilling, with line width of 10 cm and covered with 1-2 cm of soil. The seedlings were borne in 10 days (Zhijia and Xuebin, 2001).

One needs to loose the soil and remove the weeds during the germination to seedling stage, before transplantation. 37.5-75 kg of urea per ha was applied for 1-2 times in the nursery field. About 90,000-105,000 seedlings with 40 cm height and 0.5-0.8 cm of diameter, were produced in the nursery. Experts recommend 1-20 July period for sowing time of seeds for the purpose of raising of nursery from the seeds.

Pots/Trays

Seabuckthorn seeds can also be seeded indoors (green house) in sterilized soil pots, during January or early February. For better growth of seabuckthorn plants, one seedling per pot should be allowed to grow for 3 months before its transplanting in field during early April or May. Special trays with a size of 2"-4" diameter and 12" depth, manufactured for tree seedlings, can also be used for early mass production under greenhouse conditions.

Management

Within 12-18 days of sowing, seeds sprout. In case ground surface becomes dry and hard, the seeds are unable to sprout properly and seedling may curl and suffocate under the ground surface. Strong sun light may also raise soil temperature. At this time, timely cultivation of soil and breaking up the hard ground surface to avoid the effect of temperature of the rising soil on the delicate seedlings is important for their survival. Moisture level is maintained above 70-80 per cent in the beds. Seedlings may be covered with a mulch of straw to conserve the moisture or shade for protection from bright sun light (In Lu Rongsen, 1992). When the growth is about 3-4 cm, excessive dense seedlings are thinned down and weeds are pulled out. During the appearance of second pair of leaves, distance between seedlings is maintained by thinning to 3 cm and during the appearance of fourth pair of leaves to 8 cm. When the seedlings sprout 3-5 pieces of macrophylla, the seedling is likely to survive.

Irrigation

Nursery soil should be irrigated regularly to prevent drying of seeds and to improve the growth of seedlings. However, lesser irrigation may be needed during rainy days. During the rainy season (July-September), the seedlings attain a height of 5-7 cm. If the water accumulates over the seedlings, they may face mortality. In order to avoid their mortality, there should be proper draining of accumulated water, particularly in rainy days.

Weeding

One may apply herbicide in the nursery soil before sowing of seeds to control the weeds. Linuron has been found effective to control the weeds during winter. Weeding is carried out frequently to alleviate the inhibitory effect of weeds. Insecticides like gammexame or dylox are used to check the under ground insects. Efforts are being made to apply herbal pesticides.

Fertilizers and Fungicides

For a higher growth of seedlings, supplementary chemical fertilizers are applied before or after the rain or irrigation. The seeds of seabuckthorn may be dusted with TMTD at 3 g/kg seeds or the fungicide may be applied at 40 g/m² of soil before sowing. After depressive effect for 2 years due to toxicity, satisfactory plant growth in the 3rd year may be observed (Golubinskaya, 1972).

Growing in Greenhouse

The pre-treated seeds may also be sown in pots in January or early February in a vermiculite/peat moss mixture (40:60) in greenhouse conditions, with a 16 h photoperiod, at 25-27°C temperature and 70-90 per cent relative humidity. After germination within 3-10 days, and prior leaf formation, fungicide application is provided to control the seedling damping-off. Growth of seedlings may be accelerated by application of bright full-spectrum fluorescent or high pressure sodium lighting. Application of 10N, 52P and 10K fertilizers during each watering has been found growth improving for seedlings. For cultivation of plants under plastic house, the following conditions have been recommended by Russian workers (Garonovich, 2003):: fertilizers of N_{80} P_{150} K_{100} in substrate before sowing, top dressing of one year seedling with N_{100} P_{250} $K_{100,}$ liquid top-dressing with ammonia nitrate (20 g/m²), putting microfertilizers, tilling of roots before planting with IAA, IBA, esculetin during 8 hours, non-root tilling of growth points with solutions of microelements and stimulators (IAA-0.05 per cent, K-IAA- 0.02 per cent, A-NAA-0.02 per cent, etc.). While growing seedlings, one should use herbicides in doses of 30 kg/ha THA and 1-2 kg/ha.

In order to provide adequate ventilation, a moderate movement of air is provided in the greenhouse. Seedling grows fast and should be acclimated to natural conditions of light, temperature and air, before transplantation in the field. Each pot should be allowed to have 1 plant to grow for 3 months before transplantation in the field during April or May. Seedlings are transplanted in the field.

Transplantation

After 1 or 2 years of growth under nursery conditions, when the seedlings have attained the height of about 20-30 cm, and diameter of more than 4 mm, defoliated seedlings with well developed roots and nodules, are uprooted from a wet nursery soil. Seedlings are graded, sorted and stored at –2°C quickly in poly bags to avoid drying of the seedlings. They may also be stored in shipping bundles, without affecting their quality. Spring (April-May) is the best time for planting of seabuckthorn in the field. The root is buried 6-8 cm deep in the pits to encourage the development of another tier of roots.

In order to meet the demand of development of adequate seabuckthorn resources with a regulated ratio of male and female plants, propagation from cuttings should be preferred. Plants raised from the stem or branch cuttings maintain the genetic properties of mother plants and plants can bear fruit 1-2 years earlier than the plants raised from the seeds. Therefore, propagation from cutting is an important technology for raising artificial plantations and introducing and acclimating new and improved varieties of seabuckthorn. Seabuckthorn can be propagated using either hardwood or softwood cuttings, layering or suckers.

PROPAGATION FROM HARDWOOD CUTTINGS

The main advantage of propagation by hardwood cuttings is that plants can be propagated under natural as well as greenhouse conditions. It is easier to collect and transport the hard wood cuttings. Nurseries of seabuckthorn can be raiused from hardwood cuttings on small scale (Plate 2) and large scale (Plate 3). Studies on seabuckthorn reveal that cuttings take roots easily but the rate of rooting is uncertain and susceptible to environmental conditions.

Time of Collection

Cuttings may be collected before winter (October-November) or early spring (March or early April) before bud burst. Russian studies (Kuznetsov, 1995) have found that collection of cuttings before bud break, soaking in water (18-20ºC) for 7 days and planting them in the field with dark polythene mulch is beneficial.

Age of Cuttings/Plants

Age of the plant decides the quality of the cuttings, as cuttings collected from old trees does not root easily, whereas cuttings produce roots easily collected from young plants of 10-15 years. The plants of 5-12 years age can be used for the collection of cuttings.

Tools for Collection

Application of sharp scatiers, knives or pruning shears are used for harvesting cuttings from the branches. A faster collection of cuttings is carried out by pneumatic pruning shears or scatiers. The

Plate 2: Propagation of Seabuckthorn from Hardwood Cuttings

Plate 3: Raising of Seabuckthorn Nursery on Large Scale (*Source*: Lu Rongsen)

blade must be sharp for the clean cut, as rough cut does not heal quickly and often decay. Make the lower cut under the bud and upper one above the bud. We discard the non-lignified upper part of the branch (less than 0.5 cm in diameter) and divide the lignified part into 15-20 cm long cuttings.

Male and Female

Hardwood cuttings should be chosen from healthy, well developed plants in fruiting stage. Cuttings are collected during dormancy stage either during early spring (March-April) or winter (October-November) from the specially and differently marked male and female plants with desirable characteristics. Female and male cuttings are collected generally in the ratio of 9:1 respectively and properly maintained separately.

Size of Cuttings

Diameter and length of the cuttings influence the rooting rate. The cuttings are collected from the 2-3 year-old branches of the plant. 1-year old cuttings (lesser than 0.5 cm diameter) lack stored food and unable to produce roots and pretend to die. Generally cuttings of about 20 cm length and 1.0-1.5 cm in diameter are collected for propagation from hardwood cuttings (Shuhua *et al.*, 1989; Singh, 1995).

Storage During Winter

Most simple method of storage of cuttings during winter, also meeting chilling requirement of cuttings, is to make bundles of cuttings, attach a label of the variety and put them into a 1 m deep, 1-2 m wide and 1-3 m long pit, cover the pit with wet sand. Other way is make a bundle of 50-100 cuttings, dip the cut ends in paraffin wax, and store the bundles at –4°C or in a snow cache until used for planting in the nursery (In Li and Beveridge, 2003).

Treatments

Cuttings collected during early winter or spring need special treatment.

Sand Treatment

Experiments have shown that storing of cuttings in sand at low temperature improves root forming, which may be related to more or less of presence of inhibitor in cuttings. Russian workers have found that 2-year-old branch cut before bud break and stored for 10 days in wet sand at 10 to 15°C, gave 100 per cent rooting (Kondrashov and Kuimov, 1987).

Water Treatment

The base parts of cuttings should be immersed in running water for 24 hours or bucket for 1-3 days, changing its water twice a day (Shuhua *et al.*, 1989).

Hormone Treatment

To improve the rooting rate, base part of the cuttings is dipped quickly into 500 ppm NAA (Shuhua *et al.*, 1989) or IBA at 50-200 mg/litre before their plantation (Tsar'kova, 1988), which accelerates root formation in the cuttings. Good result has also been obtained achieved by treatment with Krezatsin (tris-2-Okyethylamnonium orthocresokyactate), a synthetic analogue (20 mg krezatsin/litre). Dark polythene mulch has a beneficial effect on rooting, use of transparent polythene improved the growth of transplanted planted.

Soil Media

Adequate aeration and water permeability of soil are required for vigorous rooting in hard wood cuttings. The nursery soil should be light soil sandy loam, rich in organic nutrients, with a pH of 6.5-7.5. In the year of plantation, nursery field should be prepared in the spring, use harrow on the soil, ploughing to a depth of 5-8 cm. Before planting the cuttings, 3-5 days in advance, we provide irrigation to the nursery soil, if the soil moisture is less than 70 per cent (Li and Beveridge, 2003). In case, cuttings are planted in open in a nursery, soil must be loose and with proper drainage. Adequate aeration and water permeability of soil are suitable particularly for root formation in the hardwood cutting. Under nursery conditions, a soil mixture of soil: sand: FYM with a ratio of 5:3:1 along with a small amount of soil collected from the seabuckthorn vegetation (for *Frankia* bacteria) provide better soil conditions for raising a healthy and vigorous nursery of seabuckthorn for a massive afforestation programme (Shuhua *et al.*, 1989).

Time of Plantation

The cuttings can be planted in the nursery field from early to late spring Osipov (1983). Plantation of cuttings in during late March-early April gives good results of rooting in India. Lebeda, Ukrain suggests that during the early spring, when the soil at the depth of 15-20 cm warm up to 5°C, cuttings are planted in the open soil.

Spacing

Cuttings are plated obliquely in the wind direction at spacing of 20 cm x 10-20 cm. Half part of the cuttings remains out of the soil, with 2-3 buds in the upper part of cutting. After plantation, to avoid the air drying of cuttings, soil around the cuttings is thoroughly compacted. Immediately, after plantation, cuttings are well irrigated. Cuttings may be planted in the nursery field in twin rows of 70 cm x 20 cm, at a 5-7 cm spacing in the row and at the density of 3,60,000-4,00,000 cuttings per ha. For small scale raising of nursery, cuttings can be planted in raised beds of 1 m wide in a 10 x 10 cm grid (Li and Beveridge, 2003). Cutting start producing roots on the 8th day after plantation. When complete and semi woody root system is developed with lateral roots and root nodules, then the cuttings on plants can be transplanted into nursery. After that, the second batch of cuttings can be planted. The cuttings of the second batch are also selected at the same time when selecting the first batch and stored in wet sand at low temperature before plantation (Shuhua *et al.*, 1995).

Weeding

It has been observed that rooted cuttings of *Hippophae rhamnoides* may have a poorly developed root and shoot system. One can apply the Simazine and Lenacil at 1-3 kg/ha and 1.5-4 kg/ha, respectively, which can increase the seedling height, growth, numbers of shoots and root development, as compared to weeding by hands (Shlyapnikova, 1985). Application of herbicides in 2 split dressings 7-10 days after planting and 4-6 weeks later enhanced the root establishment and seedling growth.

Biofertilzers

Use of appropriate *Frankia* bacteria and mycorhiza in seabuckthorn mycorrhizal roots shows significant improvement in plant growth, phosphate uptake and nitrogenase activity (Gardner *et al.*, 1984; Gulati *et al.*, 2001). During winter, cuttings are covered with thin layer of pine needles or fallen leaves. During next spring, they are transplanted in the field or allowed for more growth.

Irrigation

Monitoring of soil moisture is done during the growing season and moisture level of soil is maintained at 80-100 per cent by irrigating the field once a day in dry hot days, as the cuttings are developing roots. However, frequency of irrigation can be reduced during rainy days and when the root system is well developed. Then moisture level of field can be maintained at 70-80 per cent. While majority of the plants attain a reasonable height for transplantation in next year, some poorly grown plants can be held back for second year of growth under nursery conditions (In: Li and Beveridge, 2003).

Canopy Technology

In order to prevent frost damage to young plants, they are covered by grass screens and plastic film canopy (Shuhua *et al.*, 1989, 1995). Site may have the bitter cold with snow in winter, dry and windy in spring. In order to improver the environment of planting material (cutting), plastic arch canopy of 33 m long and 10 m wide is established over the nursery beds. Small arch canopies of 4.5 m long, 1.3 m wide and 0.5 m high are installed inside the plastic canopy from late March to late April. Small seedlings produced are covered with grass screens, which might prevent them from being frozen.

Pot Method

This is the most common method being practiced in India. Pot size not less than 20 cm^3 is suitable for raising plants. In the pot, loose and air-penetrable rooting media with soil mixture or long-fibral peat should be used. Cuttings of 18-20 cm length and 0.8-1.3 cm diameter with well-developed buds, are obliquely and half planted in the soil media in the poly bags (Plate 2). Before sticking, all buds except the two terminal ones should be removed from male cuttings, whereas in female cuttings, buds should be removed only from the terminal part (Bruvelis, 1991).

Growing Under Greenhouse

Seabuckthorn can also be propagated from hardwood cuttings under greenhouse conditions (Plate 4). Before plantation, cuttings are stored in sealed plastic bags at 1°C for 30 days. Cuttings can be planted in containers filled with peat moss and perlite mix (3:1, v:v).

The cuttings are planted in way that 2 buds remain above the media. Media is kept moist but not saturated. The pH of media should be higher than 6.5. For a better result of rooting, day temperature is maintained at 25°C and 5°C during night in the greenhouse. Once the shoots start growing, the photoperiod is set at 16 hrs light and 8 hrs dark. Once the shoot attains the height of 2.5-3.0 cm, we remove the shortest shoot. Rooting take place in cuttings within 15 days of plantation in the media. After the roots attain the length of 1-2 cm, the seabuckthorn plants are fertilized with a soluble fertilizer (20N, 20P, 20K) once a week. Cuttings, with roots and shoots, are maintained in the greenhouse for 40-60 days, before transplantation in the field. Rooted cuttings can be directly planted in the field, however, planting in pots under a controlled conditions for 1-2 months before transplantation, give better results (Li and Beveridge, 2003).

Over Wintering

In very cold environment, transplantation of cuttings in the field or storing indoors, is some times problematic. Therefore, the cuttings should be acclimatized and hardened off before their transplantation from the propagation beds to the field. In order to over wintering the cuttings, the simple method is the covering of cuttings with microfoam or mulched with peat moss or wood chips. Or they may be removed from the nursery beds and stored at –2°C. In another way, the well-rooted cuttings are transplanted in the field in late August, particularly in the method, where cuttings have been rooted directly in the containers (Li and Beveridge, 2003). Once transplanted in the field, the plants should be moderately irrigated until late summer. Weeding should be done but deep cultivation close to the plants is avoided.

Large Cuttings

Russians have also used a new technology of planting of long cuttings in the orchards (Kondrashov, 1994). He prepared the cuttings of 40-50 cm length of seabuckthorn and other horticultural crops and planted them vertically or at as angle of 60° into 35-40 cm deep holes, leaving 3-6 cm above the soil surface. Survival for seabuckthorn and other plants ranged from 73 to 100 per cent. However, this methods needs to be tried under different soil and climates conditions before adoption for raising orchards or plantations.

PROPAGATION FROM SOFTWOOD CUTTINGS

Propagation form softwood cuttings is widely practiced in Europe and China. The advantage of propagation from softwood cuttings is its high success rate. Massive nurseries can be raised from the green cuttings under controlled temperature and humidity conditions in mist chambers. Therefore, unlike hardwood cuttings, they are not prone to environmental changes, which ensure vigorous root formation and growth in the cuttings. The method needs artificial mist sprays and a plastic film house with other equipments. In this method, softwood cuttings (15-20 cm. long) are collected from plants, when the shoots begin to become woody. We remove the lower leaves, leaving 2-4 leaves at the tip and dip into rooting hormone before rooting in media like sand or perlite. We need to pay special attention to the moisture of the media. Rooted cuttings can be planted in pots or poly bags for 1-2 months before transplanting in the field.

Time of Collection

Cuttings should be collected from mother plants (Plate 5) during the early morning, when the seabuckthorn leaves are covered with dew. Cuttings collected during intensive growth of mother plants gave low rooting percentage, but again at the termination of active growth gave higher rooting. It has been observed that semi-hardwood cuttings, collected in late June or early July in Altai region of Siberia in Russia was beneficial (Potapov, 1978). In China, the green (semi-wood) cuttings should be taken during middle of July to the end of August. In most of the cases, cuttings are generally collected during end of June to beginning of July, when there is a beginning of lignification of one year old branches. Cuttings are collected on 3-4 mm below the buds by sharp knife or by the blade of safe rajar.

Size of Cuttings

The cuttings from small or thin young shoot lack the ability to form the roots. Cuttings from highly lignified branches also have a poor capacity to form roots. Studies have shown that the shoot with medium growing power, in the same layer of the crown, of the same age, exposed to sunshine, are most suitable for raising the nurseries of seabuckthorn. In Russia (Pletneva, 1983), small cuttings of 15-20 cm length, with 2-3 leaves, are best suited for propagation. Swedish workers have observed that the cuttings of 10-12 cm collected from the top part of a branch gave better results, than the cuttings collected from the middle and base parts of branches. Cuttings of 15-20 cm length are collected from the branches and immediately put into the plastic film bags to protect them from wind drying and withering. These cuttings can remain fresh for 3 days in this condition or 7 days in water without losing their rooting capacity.

Rooting

The rooting rate of cuttings with top bud is much higher than without top bud. It is presumed that the top bud produces hormones to stimulate rooting. Rooting rate of cuttings from 1 year old branches is higher than 2 year old and 3 years old branches. Cuttings collected from the upper part of the same plant had higher rooting than the lower part. Green or soft cuttings of seabuckthorn have stronger ability to produce several adventitious roots within a short period of plantation (Shuhua *et al.*, 1995)

Hormone Treatment

In case, we want to avoid hormone treatment, basal part of cuttings is placed into running water. However, hormone promotes root formation and healthy growth of plants. Russian expert recommend the treatment of basal part of softwood cuttings of seabuckthorn with 50 mg/l IBA and plantation in peat:sand:: 2:1 mixture (Avdeev, 1976). Other Russian workers recommended treatment of cuttings with IBA (100-200 mg/litre), IAA (200-400 mg/litre), NAA (50 mg/litre) for 16 hours. Chinese experts suggests treatment of base part with 500 ppm NAA for few seconds before plantation (Plate 6)

In China, before the treatment with hormone, cuttings are placed in water to keep them fresh. Then these cuttings are taken out of the water and bundles of 40-50 cuttings, kept together by a rubber band, are placed into a box filled with solution of 50 ppm IBA at a depth of 3-4 mm (In Lu Rongsen, 1992). All the cuttings soaked in the IBA solution are covered with a plastic film to maintain a temperature of 20-25°C for 16 hrs. Before planting, these cuttings should be taken out of the IBA solution and washed to remove the hormone. In China, it was found that a cutting dipped in NAA solution at 300-1000 ppm for 10 seconds accelerated root formation up to 80-100 per cent (Shuhua *et al.*, 1995).

Plantation

Lower part of cuttings/shoots is planted in the soil matrix of sand: saw dust: soil from seabuckthorn vegetation with a ratio of 10:3:1 or 10:7:0, which should be sterilized with 0.5 per cent $KMnO_4$ solution before putting into containers (Shuhua *et al.*, 1995). Cuttings/shoots can also be planted in plybags with vermiculite mixture (Plate 7). Seedlings can be transplanted to nursery, from containers, after 25-30 days in beds and 7 days in shed for training.

Mist Chamber

Cuttings require 80-90 per cent humidity to remain turgid during the rooting process. Therefore, the propagation from the softwood cuttings is carried out in the plastic film house, equipped with an artificial mist sprayers and drainage system. The seed bed is prepared flat and the medium of the seed bed may be a mixture of sand and soil (3:1). Cuttings are planted in the media with at least 2 leaf nodes are in the media below the surface. In China, it was observed that the rooting rate and the root biomass are the maximum, when the matrix of sand: saw dust (and soil under seabuckthorn vegetation) is prepared with a ratio of 10:7 (Shuhua *et al.*, 1995). It is sterilized with 0.5 per cent $KmnO_4$ solution. The seed bed is watered to keep the medium wet enough. The distance between the furrows should be 7 cm and cuttings 3 cm. Then the cuttings are inserted into the soil mixture at a depth of 1.5-2.0 cm. Placing of cuttings closely may be spoiled by mould. After the planting of cuttings, seed bed is again watered to make the medium around the cuttings settled down.

In Ukraine, the delves are dug up to a depth of 40 cm for the making the arrangement of glass beds, plastic film glass house or covered beds. The bottom parts are filled with layers of gravel or crushed bricks (15 cm) and layer of coarse-grained sand (5 cm), then layer of humus:sand::1:1 soil (15 cm). Upper most layer of substrate of pure coarse-grained sand or mixture of sand with peat with a ratio of 3:1 (5 cm). Russians use two layered substrate, which consisted of well composed manure, sand and soil in equal proportions, spread on the floor of the greenhouse, a sand layer made on the top of the substrate.

Humidity can be controlled in a number of ways, but the best way is with automatic misting units, such as brass mist nozzles, controlled by electronic time clocks and a solenoid. The mist bed is a clear polyethylene tunnel (or mesh shade cloth), located on 7-10 cm of coarse gravel. The basal heat is provided in the propagation bed, using a heating cable, so as to maintain the temperature of media at 25°C. Temperature in the rooting tunnel should exceed 40°C. Adequate ventilation is also provided to the cuttings. Cuttings can be rooted in any substrate or media, providing good air and water relationship. Media mixtures like peat : vermiculite : perlite mixture (50:25:25 by volume), Sunshine mix (Fisons), Premier promix 'HP' have produced good rooting and growth of plants from cuttings (In Li and Beveridge, 2003). The mist system is kept on day and night for 30 days. Initially 10 seconds of mist is supplied every 30 minutes. After the root start developing, we gradually reduce the duration of misting. The cuttings develop roots in 4-6 weeks.

Management

Aftercare improves the rooting rate and survival of cuttings and plants. The optimal air temperature for developing roots should be 22–27 °C (Garonovich, 2003). Relative humidity should not be lower than 90 per cent. During the first few days, beds are irrigated frequently (3-5 times a day), but not much, by using a water can or pipe with automized nozzle. During hot sunny days, cuttings are shaded by glass or film or pulling gauze over them. Weeds are not allowed to grow. After 15 days, watering is reduced to 1-2 times a day, but adequate. The rooting of cuttings depends upon the temperature and moisture in the matrix (soil and air around the cuttings. At the time of planting of cuttings, the temperature of the soil should be higher by 1-3 °C than the air. Generally, when the daily average temperature of open land is up to 18.5 °C, the average temperature of the soil is up to 19 °C and the relative humidity of the air is 40-50 per cent. However, within the plastic film, the average temperature of day and night should be 19 °C, the average temperature of the soil should be 24 °C and the relative humidity of the air should be 88-100 per cent by using spray-mist in time (Plate 8). This high level of humidity is indicated by the appearance of water layer on the leaf surface of the cuttings. In case, the water layer disappear, it is necessary to spray mist until the leaf surface retains the water layer again.

After the cuttings have already produced sufficient roots, the humidity should be gradually decreased and the film house should be well ventilated. With the onset of autumn, the rooting cuttings should be gradually tempered. First, the plastic film is removed for a short period every day and this is increased day by day. After 15-20 days, the plastic film can be removed permanently.

Seabuckthorn nurseries can also be raised from softwood cuttings in open field with adequate water sparying facilities (Plate 9).

Experts (Garonovich, 2003) recommend fertilizers of N80, P150 K100 in soil media before sowing, top dressing of one year old seedlings with N100 P250 K100, liquid top dressing with ammonium nitrate (20 g/m^2), putting microfertilizers, tilling of roots before planting with IAA, IBA, esculetin during 8 hours, non root tilling of growth points. Canadians reported that to ensure the vigorous plant growth under mist system, apply fertilizers in the root zone with a weekly application of 20-20-20 fertilizers at 2000 ppm (2 g/L) applying them immediately after the first roots are formed and continue during the growing season (In Li and Beveridge, 2003).

Over Wintering

After the onset of winter, it is necessary to cover the place where the roots have been developed with a layer (5-6 mm) of taller leaves in order to protect them from cold damage. During growth period, shoots may be trained to prepare them for winter. For the same, careful aeration of whole structure is carried out (but not for long, open up plastic film cover or slightly raise hotbed panes). In mid of August, plastic film covers and panes are removed completely. For over wintering and to save them from frost and snow damage, mulches of 5-6 cm thick layer of needles, leaves, dry humus or peat are spread over the beds. Under the mulches, poisoning baits are placed against the mice. Before the initiation of bud burst in next spring, rooted cuttings are dug out and transplanted on another site at 15-20 cm x 30-40 cm spacing for better growth before raising under orchard conditions. After one year of growth, seabuckthorn saplings are transplanted at wide spacing (*e.g.* 3 m x 1 m) under orchard conditions (In: Li and Beveridge, 2003).

LAYING AND SUCKERS

In Russia, air layering has also been found useful in raising nurseries of seabuckthorn. In a study on seabuckthorn, Dar Katuni variety's leaves and current year's shoots were removed from 10-20

Plate 4: Seabuckthorn Nursery in a Greenhouse (*Source*: Lu Rongsen)

Plate 5: Seabuckthorn Plants as a Source of Shoots for Propagation from Softwood Cuttings

Plate 6: Treatment of Base Part of Seabuckthorn Softwood Cuttings/Choots with 500 ppm NAA

Plate 7: Plantation of Seabuckthorn Softwood Cuttings/Shoots in Pots with Vermiculite Mixture in a Mist System

Plate 8: Watering of Seabuckthorn Shoots/Softwood Cuttings Under Mist System

Plate 9: Raising Seabuckthorn Nursery from Soft Cuttings with Spraying Equipment (*Source*: Lu Rongsen)

cm of shoot base and the shoot was ringed. The 2.5-3.0 cm wide ring was treated with IBA (17 mg/g) or IAA (20 mg/g) paste and the area was covered with Sphagnum moss and polyethylene. After 25 days, some 49 roots were found on treated shoots and only 2.8 on the control shoots (defoliated and ringed only) (A. Bruvlis in Li and Schroeder, 1996). Root cutting has also been used as an effective propagation method for seabuckthorn. Root cuttings were planted in pots and placed in greenhouse for 6 weeks before being transplanted to the field at a spacing of 8 x 20 cm. Before planting, cuttings need to be acclimated to field conditions by placing pots in a shady area for 7 days. The best results were obtained in sandy loam at pH 6 to 6.5 with medium humus content. Seabuckthorn easily produces suckers within a few years of planting; therefore, it can be a good source of propagation (Kondrashov and Kuimov, 1987; Singh and Dogra, 1996). The invasion of surrounding areas by seabuckthorn is quite possible. Therefore, routine management and herbicide application are the best control measures for this potential wildness characteristic of seabuckthorn.

GRAFTING

Seabuckthorn experts (Danilin and Grebennik, 1977) have successfully tried grafting in the male and female plants of seabuckthorn. Some of the scions, used, were treated with a solution of heteroauxin. Male rootstocks of seabuckthorn gave better scion survival and also better scion growth in the first year. The mean increment of scions treated with heteroauxin was greater than that of controls. The cuttings of 7 9 mm diameter are prepared during autumn and stored at low temperature or snow, to prevent them from drying. Cuttings can also be prepared during early spring. Grafting is carried out in one-year-old branch. The cuttings must carry unbrust buds. The place of grafting is tied with a piece of polythene sheet. After the 40 60 days, polythene sheet is relaxed or removed.

Chapter 14
Field Development

Field development includes operations of site selection, land preparation and application of fertilizers etc. before plantation of the crop (In: Lu Rongsen, 1992; Li and Beveridge, 2003).

Land Preparation

In order to improve the orchard productivity, it is always useful to carry out land preparation operations, particularly in an infertile site. The land preparation, which should be started 1-2 years in advance, includes the removal of stones, weeds, developing a planting bed, application of manure and improves the water drainage etc.

Site Selection

It must be ensured that site for the raising seabuckthorn gets enough sun shine, since seabuckthorn is a sun loving plant, although it grows in cold regions. Shade has been found to prohibit flower or fruit formation in the plants of seabuckthorn. Shade is also known to promote the pest attack on the plants of seabuckthorn. Seabuckthorn orchard is developed away from the site with big trees or other shade. It is also desirable that there is natural drainage of water on the site like gentle slope or drainage system is created by making water channels to drive out extra irrigation or rain water, otherwise standing water in the seabuckthorn stand may kill the plants. It is also necessary to know about the history of use of the site before planting seabuckthorn under nursery or orchard, *i.e.* farming practices, crops raised, or abandoned lands, application of pesticides and herbicides and their residual effects, diseases and pest attacks on the crops raised, flooding and, siltation of the site, which altogether may affect the survival and growth of seabuckthorn.

Seabuckthorn is known to grow naturally under varied soil and climatic conditions. Seabuckthorn is known to be cultivated preferably in sandy soil, however, Russian experience shows that productive plants should be raised in fertile soil with high water holding capacity like black earth type of soil. Heavy clay soils are not suitable for raising plantations.

Although seabuckthorn naturally grows on infertile sites, it is useful to carry out soil testing for physical and chemical characteristics of the site. Seabuckthorn grows faster on well drained light to medium sandy loam. Soil with clay and heavy loam without organic material, are not suitable for seabuckthorn cultivation. Seabuckthorn is also not suitable for compacted wetlands or soil prone to long time of flood. In natural conditions, seabuckthorn is found growing profusely on a wide range of soil types, but does better in soils with a light physical structure, rich in nutrient compounds, with a pH near neutral (pH 6.5-7.5). Best growth occurs in deep, well drained, sandy loam with ample organic matter. Therefore, soil salinity level less than 1.5 mS/cm has been recommended for seabuckthorn cultivation (In: Li and Beveridge, 2003).

Seabuckthorn has been cultivated on various types of soils: on chernozems, brown soils, grey forest soils, turf carbonates, turf-podzols, peaty loams and peaty-swampy soils with various physical structures such as sandy, semi-sandy, semi-clayey, and even clayey in all horizons, while keeping in mind the suitable conditions of root system regarding water and air relationships and soil pH near normal. Very light, sandy soils have low water holding capacity and also poor in mineral elements. Therefore, it requires addition of organic matter before plantation of seabuckthorn. Similarly clayey soils are not suitable with high density and water retention characteristics.

Site Preparation

Preparation of land for seabuckthorn cultivation must be started one year in advance. Site preparation operations also depend on the type of climate, soil, previous crop raised or vegetation growing etc. Good soil conditions help the plants for better growth, *i.e.* better aeration, adequate moisture, good content of organic matter, P and K and soil pH near to neutral. In heavy clay sands and light loams, addition of lime and organic manure is beneficial. In the heavy clay and loam soils, sand and manure amendments are required. The rhizomes and roots of weeds are completely removed before the plantation. In the sandy loam soil, it is useful to mix some clay or loam for the water retention. On an infertile land, some measures have to be taken to increase the organic matter content. Some of the operations are as follows:

Summer Fallow

We can cultivate the land for 1 year before planting seabuckthorn, to raise soil moisture and control the weeds also.

Green Manure Cropping

In order to increase the organic matter in the infertile soil and soil loss due to erosion by air and water, one can raise green cover of crops by sowing the seeds of barley, wheat, rye or oat at the rate of 80-150 kg/ha in the fall and plowed under in the early spring to allow decomposition before plantation of seabuckthorn. Raising of green manure crops for 2 years can produce 400-500 tons of green organic matter for the soil, which is equal to 800-1000 tons of semi-decomposed animal dung. The operation is very useful for erosion prone sites.

Subsoiling

In case, there is a hard pan in the site, which create problem in water drainage, subsoiling is done. In order to break up the hardpan layer, one can involve loosening a narrow strip of soil 50-75 deep.

Discing

Medium heavy texture soil sites, with a lot of silts and clays, are deep tilted and disced. Fall discing helps in loosening the soil, giving the better retention of moisture and tilt, which help in better spring planting of seabuckthorn. Spring discing 2-3 weeks before planting seabuckthorn, checks early germination of weeds.

Weeding

Herbicides like sodium trichloroacetate, dalapon and ammonium 2,4-D can be used to destroy the perennial weeds.

Ploughing

Hardy structure like stones and other undesirable materials are removed from the site. Till autumn, land is deeply ploughed and harrowed. The planting site is ploughed twice to remove both perennial shrubs and annuals.

Application of Sand

For the heavy soils like clay, sand is added at the rate of 10 15 kg/m^2.

Soil Tests

Soil analysis tests are always useful and the most accurate guide for application of fertilizers and lime. It is important to determine the soil fertility and pH levels before raising the orchards/ plantations so that necessary lime and fertilizer could be used to the soil.

Lime Application

Seabuckthorn is adapted to a wide range of soil pH. Soil acidity and alkalinity, except at extreme levels, are not limiting factors, however, the most favourable range is pH 5.5-7.5 for a better seabuckthorn growth. If the soil pH is too low (acid), it can be corrected by using lime like dolomitic limestone. Dolomitic limestone is recommended as it contains a significant quantity of Mg, which is an essential and often deficient plant nutrient. Quicklime, caustic lime and burned lime are not suitable for the agriculture land (Li and Beveridge, 2003).

Fertilizer Application

Seabuckthorn raised on fertile lands like grasslands, rarely requires the fertilizers. Although, seabuckthorn grows well on natural sites of marginal lands like river sides, slopes, hill tops, poor in soil nutrients and moisture, which are not fit for agriculture, however, supplementation of fertilizers particularly phosphorus at the time of planting of seabuckthorn, and irrigation provided to the crop when the fruiting start, improves the fruit yield in the seabuckthorn plants planted on poor sites of wastelands or marginal lands. Light sandy soil has poor water retention capacity, however, it may be improved by the addition of organic matter, manure etc. In sandy soil, one can use surface tillage, using a cultivator and add medium amount of well-decomposed manure. Experts (Lebeda pers. Comm.) recommend, prior to tillage, addition of organic manure (10 15 kg/m^2), phosphate (40 g/m^2) and potassium (10 15 kg/m^2) and calcareous fertilizers (300 400 g/m^2). In order to maintain adequate fertility of the ploughed land, compost or barnyard fertilizer should be spread on the land. However, in case, the content of soil humus is more than 4 per cent, the amount of organic fertilizer can be lowered by 50 per cent. Green manure may also be a substitute to organic fertilizer, in the absence of the latter. For an acidic soil, lime should be added to soil to increase the availability of minerals in the soil. Further application of 600-800 kg/ha of calcium superphosphte in the soil, before plouging, is useful (In Lu Rongsen, 1992).

Chapter 15
Plantation Methods

Seabuckthorn orchard may stay productive for long years of 15 to 20 years; provided we apply appropriate plantation methods, *i.e.* time of plantation, appropriate ratio and arrangement of male and female plants, right row orientation and spacing and plantation methods, which will ensure the plants are well established and attain better growth.

Time of Plantation

High survival rates of seabuckthorn plants are obtained, when they are dug out from the nurseries and immediately transplanted in the field during early spring (late March to early April), before the sprouting of the leaves. Seedlings or rooted cuttings are planted in the dormant stage. Planting at this time minimizes stress and takes advantage of new growth following dormancy. If possible, transplant the plants during the cool, cloudy, calm days with minimum wind.

Generally, plantation of seabuckthorn is recommended during early spring (late March-early April). However, this is generally recommended for private and small plantations, where regular irrigation after plantation is assured. Lack of irrigation during summer (May-June) may cause mortality of plants. In order to carry out large-scale plantations, best time is autumn (October), because before that we can carry out large-scale operations of land development, plants collection from nurseries, transport, storage of saplings and adequate time for plantations in a scientific way. In Altai area of Russia, plantation of seabuckthorn is recommended during 15-20 days before the start of the frosts, during early and mid October. However, in Northern provinces of China, farmers do plantation during spring period.

Quality of Nursery Material

For better survival and growth of seabuckthorn, we need to have quality plant material under nursery conditions. Two-year old saplings are preferred for plantation in the field than one-year-old saplings. The standard 2-year old saplings should have a stem diameter not less than 15 mm for the first grade plant material and 8-10 mm for the second grade plant material. The height of the saplings should be not less than 50-60 cm and 35-50 cm respectively. The length of roots should be 20-25 cm with 4 branching and 15-20 cm with 3 branching, respectively. One-year old standard saplings should be not less than 30 cm in height. One and 2-years old saplings are those plants which have grown for 1 and 2 seasons respectively under nursery conditions.

Hardening Off

This process is required in case we carry out plantation of rooted softwood cuttings in August. Non-dormant plants are best planted before a period of growth, so that root system could develop. Before transplanting the non-dormant plants, hardening-off is required (Li and Beverdige, 2003). The hardening-off process help the plants to gradually adapt to field conditions by slowly lowering temperature (adapting) and creating some moisture stress. Planting at this time allows some time for root growth, but also provides enough time for natural hardening.

Pruning of Roots

Seabuckthorn saplings can be collected from the nursery before winter and stored at low temperature (2-4°C) for storage during winter and plantation in early spring. In order to separate the damaged root during harvesting or diseased root during storage, pruning of roots is carried out before plantation. Roots may be cut to a length of 10-15 cm length for vigorous rooting in the course root (Li and Beverdige, 2003). Seabuckthorn roots are kept shaded, wet and protected from sun or air-drying during storage, transport and transplantation.

Transportation

Moist the nursery field 1-2 days in advance before collecting the plants for transplantation. We dig out the plants, never pull out them, other wise it will damage the roots. The time of interval of the digging out and transplantation should be maintained as less as possible for the higher survival of the plants. During the transportation from one place to another, the seedlings should be protecting from drying by wet cloths or spaying water on the roots etc.

Male and Female Ratio

Seabuckthorn is a dioecious and cross pollinated plant, where pollination takes place by wind. Therefore, adequate number of pollinators is required for the maximum fruiting in female plants. Russian experts have found that 6-7 per cent male plants are enough for adequate pollination, whereas, German experts have recommended 8-12 per cent male plants (Albrecht *et al.*, 1984; Wolf and Wegert, 1993). Russian experts recommend one male: female mixed row for every two rows of female plants. In the mixed row, every fifth plant is a male, which means, there should be 6-7 per cent male plants of the total (Gakov, 1980). Generally, the distance within which the female plant can be pollinated is about 100 m. Investigations have shown that as the distance from the female plant to the male plant (polliniser) increases (64 m or so), the yield of the female plant decreases. Therefore, for balanced pollination, the disposition of pollinisers given above is suitable.

For higher fruit yield in female plants, male variety should be cold tolerant, have a long flowering period, produce high number of pollens and grows vigorously, since, there is an important effect of characteristics of the male plants on the fruit size, yield, flavour and ripening in female plants. Male plants may also be planted on the boundaries around the rows of the female plants, as heights of the male plants are kept higher than the female plants.

Figure 4: Recommended Design of an Orchard of Seabuckthorn

×	0	0	0	×	0	0	0	×	0	0	0
0	0	0	0	0	0	0	0	0	0	0	0
0	0	0	0	0	0	0	0	0	0	0	0
0	0	×	0	0	0	×	0	0	0	×	0
0	0	0	0	0	0	0	0	0	0	0	0
0	0	0	0	0	0	0	0	0	0	0	0
×	0	0	0	×	0	0	0	×	0	0	0
0	0	0	0	0	0	0	0	0	0	0	0
0	0	0	0	0	0	0	0	0	0	0	0
0	0	×	0	0	0	×	0	0	0	×	0
0	0	0	0	0	0	0	0	0	0	0	0
0	0	0	0	0	0	0	0	0	0	0	0

×: Male; 0: Female.

If seedlings of unknown sex are planted, once plants are flowering or fruiting, so we know sex of the plants, then remove the male plants and replace them with female plants which have been vegetative propagated from mature plants of known sex. Nursery rows are also planted near the orchard at the same time the orchard is established, and the females used for replacement of male plants (Li and Beveridge, 2003). Canadian studies have found in an orchard, planting with 4,000 trees per hectare and a 1:6 male and female ratio, should yield approximately 10 tones of berries.

Direction of Orchard

Seabuckthorn does not tolerate shade and it is sun-loving plant, hence, adequate light for growth and fruit production (Beldean and Leathu, 1985). Therefore, in order to maximize the penetration of light in the plant, plant the rows in a north-south direction.

Spacing

The space between the rows should be sufficient to allow access of adequate light to the plants and easy movement of mechanized equipments for tillage, pest control, pruning and harvesting. Therefore, Canadian experts (Li and Beverdige, 2003) recommend the distance between rows should be 5-6 m, which also allow sufficient space for proper shrub development. The distance between plants in a row may be maintained to 1.0 or 1.5 m. German experts have found that the spacing of 1 m between plants and 4.0-4.5 m between rows is useful so as to allow the use of equipments, with raising rows in north-south direction for providing maximum light. A closer spacing of 3.0 m x 0.8 m produced a maximum of 26.2 tons of fruit yield per ha obtained in 'Chuiskaya' variety at Lisavenko Institute of Horticulture, Siberia in Russia (Khabarov, 2003). In a Chinese study, highest fruit yield was obtained at a spacing of 3 m x 1 m (In Lu Rongsen, 1994). For raising shelterbelts, 2 years old seedlings are planted at 1 to 2 m, for the distance between plants within the row and 5 m between rows (Schroeder and Yao, 1995). Therefore, most experts agree for a distant spacing of 4-5 m between rows and 1.5 m between plants for better conditions of growth. However, we can lower the distance to 3 m between rows and 1 m between plants, even lesser spacing (*e.g.* 2.5 m × 0.8 m), in case we want to get maximum out put in a given land area.

A close spacing allows for higher early yields and optimise the land utilization, whereas wider spacing provide better orchard ventilation and reduce possibility of spreading pests and disease. More distance between rows makes sucker control and using mechanized equipment easier, whereas higher plant-to-plant distance of 1.5 metres facilitates pruning and mechanical harvesting of fruits from the plants. The number of plants per hectare for different plant spacing is shown in Table 2.

Table 2: Number of Plants per ha for Different Spacings

Distance Between Rows	*Distance Between Plants*	*Number of Plants/ha*
3 m	1.0 m	3333
3 m	1.5 m	2222
4 m	1.0 m	2500
4 m	1.5 m	1666
5 m	1.0 m	2000
5 m	1.5 m	1333
6 m	1.0 m	1666
6 m	1.5 m	1111

Pre-treatment with Water

As seabuckthorn roots are sensitive to drying by air and sun light. Seabuckthorn roots should never be allowed to dry even slightly. It is also advisable to place them in water for a while before plantation in the field. Bare root seedlings and rooted cuttings benefit from placing them in water; however placing roots in water for more than four hours may damage the seedlings. All trees, except those needed for immediate planting, should be kept shaded and wet. While waiting to be transplanted, plant roots need to be covered all time.

Pruning of Roots

Seedlings can be harvested before winter and stored at low temperature (2-4ºC) for storage during winter and plantation in early spring. In order to separate the damaged root during harvesting or diseased root during storage, pruning of roots is carried out before plantation. Roots may be cut to a length of 10-15 cm length for vigorous rooting in the course root (Li and Beverdige, 2003). Seabuckthorn roots are kept shaded, wet and protected from sun or air drying during storage, transport and transplantation.

Pits

We dig pits for seabuckthorn plants slightly deeper than the depth they grew in the nursery or propagation container. If they are planted too shallow, the roots will dry up. One needs to take precaution for plants in containers, in case, the rooting medium becomes exposed, and the plants dry up very quickly. Planting holes must be large enough to accommodate the root mass. A sharp spade works well; whereas the augers can glaze the walls of the hole, which is inhibiting water movement and the penetration of roots. We insert shovel vertically with blade reversed; push handle forward, then pull soil back and out of hole. The planting pits of 35–40 cm depth and 40–50 cm diameter are dug during early winter (September-November) or early spring (march-April), better one year in advance before transplanting of the seedlings. In the wastelands, some experts suggest 60 cm x 40 cm size pits in the soil. Russian experts (Dolgacheva and Aksenova, 2003) suggests 0.7 m deep pits. On sites with out any irrigation, large holes of 100 cm x 100 cm x 50 cm size can be made for acculülation of moisture for plantation of seabuckthorn plants (Plate 10). For dry sites like wastelands, a furrow

Plate 10: Digging of Holes for Plantation of Seabuckthorn on Marginal Land (*Source*: Lu Rongsen)

of 30-50 m can be made around the pit, in which 2 3 buckets of water is poured. This helps in maintaining the adequate moisture for the plants during dry days.

Soil Mixture

Seabuckthorn is a nitrogen fixing plant and does not require nitrogen fertilizers. To fill up the pit and for a better growth of plants, a soil mixture of soil: sand: FYM (manure)::1:1:1 may be used for infertile soil (Lebeda, per.comm.). Along with soil mixture, we can mix up properly double superphosphate (200 g/pit). In the light sandy soil, sand is not used. In case soil is not fertile, fertilizer should be mixed in the soil and avoid direct contact with roots of the plant.

Transplantation

Manual

Transplantation by hands gives a care to individual plant. We should not plant the seedlings in a pit, when soil moisture is below the wilting point or above field capacity. Insert the plant at a proper depth in a pit. Roots of plants are placed and spread at the bottom of the pits, while placing plant roots at proper places and regular intervals. Fill up the pits with loose soil mixture, leaving one third cavity or pit from top for water or moisture accumulation. Push the soil around the plant to avoid drying up of the plant. We pour the water into the pit. After the water is absorbed by the pit soil, this small cavity may be covered with small layer of soil mixture, humus, peat or river sand.

Transplantation can also be carried out in furrows. Manual method of planting of 1-year old sapling is carried out in 20 cm deep furrows. For 2-year-old plants, we make 30 cm deep furrows. During transplantation, 3 workers, one of which goes over the furrows, placing the saplings in vertical position in the furrow and two other workers strewing it by ground. Therefore, planting of seabuckthorn in land of 100 ha will demand about 230-250 man-days of labour.

Planting Machine

Transplantation of seabuckthorn saplings by machine needs a tractor operator and two planters. Planting is carried out by placing the seedling in the ground between the trencher blades, well ahead of the soil falling into the trench. The sapling is leaned, a little backward with the planter's knuckles, touching the ground. The planter follows through until the forward motion of the machine allows the soil to counteract any influence the packer wheels might have to slant the trees forward. Planting rate depends on soil tilt and topography, size of seedlings, planter experience and tree spacing.

Russian experts have developed a machine for planting of seabuckthorn. Model MPS-1 of this machine is being used for the mechanized planting of small-berries, fruit and flower plants. This machine can plant 500-1800 saplings per hour in 0.5-1.0 ha, with a plant to plant distance not less 25 cm and inter-row spacing not less than 1.8 m. These days, new models of forest planting machines are available in the Russian market. For example, SLN-1, place saplings in a furrow, while moving over a furrow. Furrows in the field are prepared by ploughshare. This machine plant saplings in 0.25-0.6 ha per hour, plant to plant spacing being 0.5, 0.75 and 1.0 m, whereas depth of furrow is 25-27 cm. There is another model LMD-1, which is more productive, as it can plant up to 4000 saplings per hour in a land of 3 ha per 1. But this model is used only for small saplings with high inter row density. For higher spacing of 1.5 to 2.0 m, this machine can plant saplings in no more than 1 hectare per hour. There is another model MLU-1À, which has been released recently.

In order to run the machine, 2 workers are used; however for convenient planting no less than two additional persons are required for preparing mash and delivering wetted seedlings to the machine. Therefore, it requires 50-60 man-days per 100 hectares, which is 8-10 times lesser than the manual planting. Any way of planting requires placement of saplings in a vertical position. For the development of a sprawling crown, 30 per cent cutting of the leading shoots of young plants is more useful. In this case, there is intensive growth of lateral shoots and development of a productive bush. Cutting is carried out during the next years, but it is a sanitary character and as a rule does not carried out in the large areas.

Cost of Planting Material

Depending on the spacing between plants and their rows, number of plants required for a plantation in 100 ha varies from 1,25,000-1,60,000. Cost of 2 year old planting material for 100 ha land will cost in Russia about $ US 100.000. However, in order to reduce the cost, it is advisable to purchase the 1-year-old planting material.

Chapter 16
Management Practices

IRRIGATION

Seabuckthorn can tolerate a little drought but it is a moisture sensitive plant especially in the spring when plants are flowering and young fruits are beginning to develop. For plantation of seabuckthorn in arid or semi-arid regions, irrigation must be supplied to the young plants to establish, particularly where rainfall is less than 400 mm. It does not tolerate standing water above the crown, high water table or long-term inundation (15 days); therefore planting seabuckthorn plants on sandy loam soil on slight slopes with good drainage is most suitable. The optimal soil moisture for seabuckthorn is about 70 per cent, whereas inadequate soil moisture causes a reduction in leaf area and fruit set (In Lu Rongsen, 1992).

While seabuckthorn is well adapted to dry habitats, it certainly attains better growth on application of irrigation to seabuckthorn stands, growing especially on the dry sites. We provide irrigation to the plants regularly during dry days of summer; however, reduce the frequency during the rainy days or during early winter. Dry conditions and over watering kills the plants. Watering the furrow around the pit maintains moisture around the plant and helps the better growth under dry conditions. Standing water or flooding the plants or field also kills the plants. Drainage of excess of water is required under such conditions.

Irrigation especially during first 2 years help in the establishment the young plants in dry conditions. Young plants may be sensitive to dry conditions of the site especially during the period of flowering and fruiting (April-September). The amount of irrigation to be provided depends on the local climate conditions, soil conditions (water holding capacity) stage of plant development and plant age.

The minimum soil moisture level required for growing seabuckthorn is 70 per cent for the medium clay loam, 80 per cent for heavy clay, 60 per cent for slightly sandy soil and 60-70 per cent for the sandy-loam (In Lu Rongsen, 1992). Therefore, when the moisture level in the soil is lower than above levels, particularly during the flowering and fruit development phases, irrigation must be provided. In the area of non-chernozem soil, the site is provided irrigation 3 times during the growth period (each time 400-500 m^3/ha). It was also observed in the areas of east Siberia and Lake Baykal, Russia, that when there was not adequate rain and snowfall, irrigation was found important before the onset of winter, as adequate water not only increases the soil temperature, where roots are distributed, but also provides the moisture that plant requires in the next spring. About 500-600 m^3/ha water was soaked up to the depth of 70-80 cm and was found enough for these areas. It was concluded that irrigation is very important during dry conditions, especially during flowering and fruit development phases (In Lu Rongsen, 1992).

In general, supplemental irrigation should be given to an orchard, when its soil moisture level is below 60 per cent of the field capacity. Soil moisture sensors can be used as an effective method for monitoring the soil moisture level and precise irrigation schedule. In most cases, water is applied when the soil moisture sensors show the value in the range of 30 to 60 centibars. The lower end of the range is for hot, dry climates and lighter soils; whereas the upper range is for cool, humid climates and heavier soils. Precise moisture requirements of seabuckthorn are being estimated. Excess soil moisture may be a serious problem for seabuckthorn plants especially in poorly drained soils. Seabuckthorn plants do not tolerate waterlogged soil (Li and Beveridge, 2003).

Besides water quantity, quality of water is also an important factor to consider when planning an irrigation system. Generally, surface water like river, canal, lake, reservoir or a farm dug out is usually superior as a water source than groundwater like a well, since surface water is warmer and has lesser soluble solids, especially salts. It is useful to know the concentration and composition of total dissolved solids (TDS). Surface water is generally low in dissolved solids than groundwater. A water source with a concentration of TDS higher than 800 ppm is not considered suitable for irrigation of an orchard. The salt level and in particular the amount of sodium in the water is important, while using a ground water for irrigation. Therefore, when water from a well is to be used, then it is desirable to know the salinity level, which should not be less than 1 mS/cm (Li and Beveridge, 2003).

Irrigation can be provided in various ways; however, trickle system distributes the adequate water to individual plants, with little wastage due to evaporation or runoff. Trickle irrigation is generally a method of choice of orchard growers. A good quality trickle system will last for many years. The main benefits of this system are low cost, low labour requirements, efficient water use and no erosion problems. The cost of a trickle irrigation system ranges from $1000 to $2000 per acre (Li and Beveridge, 2003). This includes pumps, pressure regulators, filters, headers, pipes, valves and emitters.

In water holding and fertile soils, there is enough 2-3 watering during the most important periods. However it is obvious, that watering on large scales is a very expensive action, and only rich enterprises may afford to organize the irrigation system.

FERTILIZERS

While seabuckthorn can grow naturally on poor soils like mountainous wastelands and riversides, however, it also requires adequate soil nutrients to produce high yield of quality fruit. During the spring, when seabuckthorn starts to develop fruits, it begins to use the soil nutrients, which were accumulated previous year. During the first phase of growing period, seabuckthorn requires a good supply of nitrogen. The nutrients are absorbed to form roots, branches and leaves. During the later stage of plant development, when stems stops growing, soil nutrients are required for the development of fruits. The nutrients are then transported gradually from the leaves to main trunk, branches and roots.

The exact nutrient levels required for a seabuckthorn orchard have not been yet established. It is important to determine soil fertility and nutrient levels of a site before planting seabuckthorn so that necessary lime and fertilizer could be applied to the soil. Seabuckthorn, just like any other crop, requires adequate soil nutrients for a high yield of good quality fruit. Seabuckthorn is capable of fixing atmospheric N by its root's nodules, however, slowly released nitrogen applied just after planting is beneficial for the better growth of seabuckthorn plants. It also responds well to phosphorus fertilizer, especially in soils low in phosphorus.

There are various ways of fertilizer application. It can be broadcasted on the soil surface and mixed into the soil using the tillage or fertilizer can be top dressed when seabuckthorn is growing. Seabuckthorn plants often require phosphate and potash fertilizers. From the third year of growth onwards, young plants are given fertilizers like ammonium nitrate in April and phosphate and potassium in autumn, after the harvest of the crop. However, excessive nitrogen application will induce vigorous vegetative growth. High vigour may increase the fruit yield up to a point, but eventually shoot growth becomes competitive with fruit set and results in a reduction of fruit yield. Seabuckthorn responds well to P application in the deficient soil. Application of calcium super phosphate at the rate of 600-800 kg/ha, in the furrows has been found very effective in promoting growth of seabuckthorn (A. Bruvelis in Li and Schroeder, 1996). In Russia, it was found that a single spring application of 100 kg N: 200 kg P_20_5: 100 kg KO_2/ha to seabuckthorn plants, considerably increased fruit size and yield (Garanovich *et al.*, 1983). Top dressing had a generally favourable effect on the plants, increasing shoot growth. Significant changes also occurred in fruit composition. In several cultivars of seabuckthorn, contents of oil and vitamin C increased and titrable acidity decreased.

Utility of application of fertilizers increases under assured irrigation. In a black calcareous soil in Siberia, application of 60 kg/ha, each of N, P_2P_5 and K_2O increased fruit yield by 23 per cent (Predeina, 1987). Russians observed an increase in the fruit weight by 9.1-34.5 per cent but decrease in fruit oil content in the trials on

Maslichnaya variety of seabuckthorn, sprayed with solutions of Cu, Mo, Mn, I, B, Co or Zn (Mishulina, 1976). The treatments also augmented fruit vitamin C content, by 32.4 per cent with I, by 30 per cent with B, and by 28.1 per cent with Co or Zn. Cu increased carotene content by 14.8 per cent and Mn by 25.4 per cent. In China, it was found that application of urea was better than compound fertilizer. Fruit yield and height increased by 20-70 per cent and 9-38 per cent per tree, respectively, by using urea as compared to control (Xinxue, 1992). They found that fertilizer application in sandy soil, lacking nutrients, promoted growth and fruit yield. However, some experts observed that application of nitrogen fertilizer might adversely affect the root nodulation. Similar observations with other nitrogen fixing plants have also been reported. Russian experts have observed in young plants that nitrogen fertilizer stimulated growth of seabuckthorn plants and their fruit yield, however, it led to decrease in resistance to *Verticillium* wilt (Tyurikov, 1993). Whereas application of potassium did not affect plant growth and yield, but increased their resistance to diseases. The experts have advised the soil analysis tests before applying the fertilizers.

Manure

Manure or compost supplies nutrients to a plant for a longer period of time. Cattle and poultry manures are commonly used in seabuckthorn orchard. Cattle manure can be used for a maximum amount of 45 t/ha and poultry manure not more than 20 t/ha (Li and Beveridge, 2003). Chinese experts have recommended the application of 45-50 tons/ha of organic manure for applying in seabuckthorn plantations every 3-4 years (Lu Rongsen (1992). Russian experts have found the better growth was also obtained by using peat compost at the rate of 60 tons/ha and 50 kg/ha, each of N, P_2P_5 and K_2O (Martemyanov and Khromova, 1985).

Lime

Seabuckthorn can tolerate a wide range of soil pH, but a pH level between 5.5-7.0 is ideal. Soil acidity can be corrected by the application of lime. Use of some dolomitic limestone is recommended since it contains a significant quantity of magnesium which is an essential and often deficient plant nutrient. Quick lime, caustic lime, and burned lime are not recommended on agricultural land.

Mineral Deficiencies and their Treatments

There is not much information available in the literature regarding the mineral deficiencies in seabuckthorn, however, we can follow the general guidelines, mentioned below in assessing whether deficiencies are occurring or not (In: Li and Beverdige, 2003):

N and/or P Deficiency

In this deficiency, leaf is pale green or yellowish and leaf senescence and dehiscence may increase. Plants are smaller and leaf area may decrease. If these deficiencies appear, soil tests need to be carried out and if necessary N fertilizer should be provided as ammonium nitrate at 20 g/m^2 early in the spring. P fertilizer (superphosphate) should be provided at a rate of 20-30 g/m^2 in the October month.

K Deficiency

In this deficiency, leaf colour is pale, there is marginal chlorosis and/or necrosis, shortened stem internodes or there is the death of the terminal bud. K fertilizer should be given at the rate of 20-25 g/m^2 potassium chloride along with P.

Mg Deficiency

While, the terminal leaves may appear normal, however, the basal leaves may have marginal chlorosis with a V pattern. This deficiency occurs when K is high or soil is acidic, especially in young and vigorously growing plants. Defoliation of leaves starts from the base of shoots and progressively affects the leaves above.

Z Deficiency

There is a delayed opening of flower and leaf buds in the spring. Plants have small chlorotic leaves, shortened internodes and small leaves along the shoot. Plant shows reduced growth and bear small fruits.

Fe Deficiency

There is a loss of chlorophyll and leaves become chlorotic. The interveinal areas of leaf become yellow, but the veins remain green. These symptoms develop first on young leaves. Fe deficiency may appear, when there is deficiency of sufficient Fe in the soil or if there is sufficient Fe but it is not available to the plants due to soil conditions (*i.e.,* high pH) or available Fe is not properly utilized in the plant. Application of iron chelates to the soil can alleviate these problems and the plants may gain normal growth.

Other Factors

Shade Effect

Shade of other trees may cause small crowns and low fruit yields in seabuckthorn. Therefore, seabuckthorn need to be raised in a sunny site with out any shade.

Standing Water

Standing water, too much water, high water table in an orchard may cause the poor growth and stems rot at the soil surface. One need to grow seabuckthorn on sloppy land or site with good drainage of water.

Dry Soil

Soil moisture deficiency results in the formation of small leaves or may cause leaves wilting, which turn yellow and drop. Seabuckthorn requires adequate irrigation especially during the dry days of the summer for a better plant growth and fruit development.

Winter or Frost Damage

Winter or frost damage may result in branch die back with late emerging and formation of the small leaves.

Heavy Snow

Heavy snow may result in branch breakage of seabuckthorn plants.

Poor Pollination

Poor pollination results in low fruit yields of seabuckthorn. One need to ensure that there are adequate numbers of male plants planted in the orchard and that male flowering is compatible with female flowering. Male cultivars with longer period of flowering and high number of pollens are preferred for better results.

Herbicide

A release or drift of herbicide from adjacent agricultural fields may result in the formation of twisted and curled leaves.

WEEDING

Control of weeds in seabuckthorn plantations is very important especially for the healthy growth of the young plants of the crop. Many weeds have a more vigorous root system than seabuckthorn and grow faster, as they compete for soil moisture and nutrients with the main crop. Lack of adequate weed control causes more seedling mortality than any other single cause. They affect the seabuckthorn growth, lower their fruit yield and make the crop some times unprofitable. Therefore, weed control is necessary for both survival and growth, especially with newly planted seabuckthorn. Weed control in an orchard is necessary unless the seabuckthorn plants are large enough in 4-5 years to shade out the weeds. Manual weeding by hands is generally practiced in small nurseries of seabuckthorn.

There are several methods to control the weeds in a seabuckthorn stand, depending on type of orchard operation, type of weeds present, soil type, and/or equipment available for the operation.

Soil management in a seabuckthorn orchard is very important especially for a new plantation, where weak plants may not be able to compete with weeds for soil moisture and elements. Manual weeding by hands may be very labour expensive and time consuming, therefore, economically unviable. Alternative methods of weed control include grass raising, green manure crop sowing, mulching of stripes and application of herbicides. However, full grass raising may also cause a strong competition for soil moisture and minerals. Alternating grass raising with seabuckthorn is more useful, however it also demands manual weeding.

During weeding operations, soil below the depth of 8-10 cm should not be disturbed, as shallow root system lie at this depth (Gonchar and Saban, 1986). In an orchard, a mechanical or hand cultivation is very effective in controlling weeds between the two tree rows. As the root system of seabuckthorn is shallow and distributed in surface soil layer, therefore, the cultivation of soil should be shallow (<8 cm deep) for avoiding the damage of its root system (Li and Beveridge, 2003). During cultivation of soil, one needs to avoid physical damage to the plants of seabuckthorn. We may provide 3 cultivations to the soil during the growing season for adequately controlling the weeds in an orchard of seabuckthorn. Mechanical cultivation may be accomplished by hand cultivation or cultivation while using specialized in-row cultivating machine. Hand hoeing, although a most time-consuming weed control measure is also the safest and the most effective method for weed control in a seabuckthorn stand.

Plastic Mulch

Application of plastic mulch is now a popular, simple and cost effective method for checking the growth of weeds in seabuckthorn. Plastic mulch improves the survival and growth of seedlings of a crop by providing a better environment, which is weed free and with adequate moisture and nutrients. Black plastic suppresses the growth of weeds mainly by blocking light necessary for the photosynthesis in weed plants and causing to some extent mechanical impeding in their growth. Input of labour and herbicides required with weed control are also significantly decreased (In Li and Beveridge, 2003). Plastic mulch is an ultraviolet-stabilized (UV) black polypropylene or polyethylene. Preparation of plastic mulch requires modification in a sheet of plastic. One needs to make some preparations in an orchard before an application of plastic mulch. Good results are obtained if the soil has been cultivated to a depth of 15 to 20 cm before the planting of the seabuckthorn seedlings. In case of a machine planting of the crop, the planting furrows should be leveled to surface for using the mulch.

We can place the plastic mulch immediately after planting the seedlings in the field. The applicator buries the edges of the plastic, securing and keeping it tightly fitted over the soil surface. An X-slit is made in the plastic at each seedling. The seedling is then pulled through the slit. Making an X in the plastic rather than a single slit reduces the chance of bark abrasion on the seedling. One needs to pull the seedlings of the crop through the plastic immediately after it is laid. Seedlings may burn and die during the hot days if left under plastic for even 5 to 10 minutes. We secure the plastic at each seedling, using a rock or wire staple. This ensures plastic is in contact with the soil around the seedling, which will help to direct moisture to the seedling and limits weed growth at the opening (In Li and Beveridge, 2003). Placing of a rubber plate around the plant also helps in controlling growth of weedsespecially near seabuckthorn plants in the plantation (Plate 11).

Plate 11: Placement of Rubber Plates around Seabuckthorn Plants for Weed Control

We can remove the weeds by hands or herbicides, which are growing along edges of the mulch; therefore do not use rototillers or cultivators, otherwise plastic will be easily dislodged or damaged. Weeds growing around the seedlings of plants of crop may have to be removed by hands. Grass can be sown between the tree rows, which helps to keep mulch in place and reduces the growth of weeds. Grass can be mowed right up to the mulch. Grass species, which are non-competitive and low maintenance, for example sheep's fescue, are ideal choices for growing between the rows. Plastic will also inhibit the suckering of seabuckthorn, therefore, preventing the development of a continuous hedge of it. Although, a little maintenance of plastic is required under normal conditions, however, the damaged or ripped plastic mulch should be repaired or secured to the soil, using staples or soil.

Rodents like mice and pocket gophers may cause a problem under mulch, if their numbers are high. Under these conditions, in order to minimize the damage, keep the grass mowed or remove weeds adjacent to the plastic and place poison bait stations under the plastic mulch. The proper installation of mulch has a significant effect on the mulch durability. In case it is not properly installed, *i.e.*

it is not anchored properly, then it will not be as effective or last for long. Wild or domestic animal, puncturing plastic or hail damage can reduce the longevity of plastic mulches. Under the normal conditions, as it is made of UV inhibitors, the breakdown and deterioration of plastic may not occur for several years (In Li and Beveridge, 2003).

Mulching of Straw

One effective method of weed control is by mulching of straw at the rate of 4.2 kg per one linear meter. Besides not allowing the occurrence of weeds, there is an enhancement of moisture capacity under straw. However, it requires the application of huge amount of 1000 tones straw for 100 hectare of seabuckthorn plantation. Therefore, it is only applicable for small plantations.

Grass Mulch

In dry conditions, where it is difficult to maintain adequate soil moisture, application of grass mulch has been found useful. Chinese experts have observed that grass mulching below the canopy of seabuckthorn plants could increase the moisture content by 164 per cent, tree height by 25.5 per cent, and crown width by 33.8 per cent and fruit yield by 108.5 per cent (Xinxue, 1992).

Grass Cover

Growing the grass between rows of seabuckthorn is an alternative to the mechanical cultivation or application of mulch or herbicides. It provides a clean surface and easy access for equipment following rain. It also checks the soil erosion in the plantation, especially if applied with mulches. The main disadvantage is that the grass needs mowing and if not controlled, it may compete with the fruit crop for the moisture and nutrients.

Grass should not be allowed to grow near the tree row. In case mulch is also used, then the grass can be allowed to grow along the edges of the mulch. However, for using herbicides or mechanical weed control, an area of approximately 60 cm on each side of the row should remain clean. Selection of a grass species is very important, when mowing between orchard rows. The grass species should be hardy, non-competitive with fruit crop and drought tolerant. It should form a resilient carpet and capable of withstanding vehicle and equipment movement during the rainy season. Height of a grass canopy is important, as a tall vigorous grass that requires regular mowing is not desirable. Generally, bunch grasses are less competitive than the rhizotamous grasses. Some grass species have been identified, which are adapted for application as a grass cover in seabuckthorn orchards. For example, sheep fescue and hard fescue are densely tufted, low growing bunch grasses which are shade tolerant. They are highly drought tolerance and have winter hardiness. They have a maximum height of 30-45 cm, if left unmowed and they only require mowing twice/year. Grass covers can be developed in the orchard in the late summer or spring. The seeds of grass is broadcast on a firm seedbed between tree rows at a rate of 45 kg/ha and then lightly harrowed. It is important that the seeds are lying in the soil not deeper than 1 cm. Once the grass is established, over seeding may be required to fill in blank spots (In Li and Beveridge, 2003).

Green Manure Crops

As alternative, sowing of the green manure crops between seabuckthorn rows is also useful especially in moist sites. This helps in improvement of soil properties and fertility and decreasing the soil erosion, while checking the weed spread. We can sow the green manure crops at the end of June, and cutting and spreading the plant manure at the time of budding as the most economically viable way of soil management both in young and old seabuckthorn plantations.

Herbicides

Considering that seabuckthorn can't compete with weeds and hand weeding is labour expensive, therefore, we need to find highly productive and suitable methods to remove the weeds. Effects of herbicides on seabuckthorn, and their residues in berries and leaves are not properly investigated, therefore should be applied restrictedly. However, some preliminary studies on some herbicides show high efficiency of their application in seabuckthorn plants. Various herbicides are applied in seabuckthorn plantation, depending on the specific variety of weeds growing there. Application of herbicides, although considered dangerous for cultivars, may be required against perennial weeds in large nurseries and orchards of seabuckthorn. Land preparation before planting is useful for successful elimination of weeds. We should pay a great attention to the killing of perennial weeds. Therefore, combination of mechanical soil treatments with herbicides application during land preparation for the establishment of seabuckthorn nursery or orchard is a very important element of intensive technologies.

The selection of herbicide to be used depends on the site conditions, soil type, climate of the area and types of weeds. Seeds of annual weeds germinate usually in the top 5 cm of the soil; therefore, the herbicides applied to the soil inhibit the seed germination and plant development of such weeds. The herbicides may be applied as pre-emergent of weeds or incorporated into the soil. Pre-emergent herbicides are applied to the soil surface and rainfall is important to move the herbicide into the soil for its effective use in weed control. Whereas, the soil incorporated herbicides are mixed manually after being applied to the soil surface. Post-emergent herbicides are applied to the young small seedlings of weeds growing vigorously. Leaves of these small weeds absorb the herbicides, which are translocated throughout the plant. These herbicides can be used to control the weeds from late spring to early fall, depending on the type of weeds growing in the field. Herbicides recommended for use in seabuckthorn shelterbelts are trifluralin, metribuzin, sethoxydim, linuron and roundup (In: Li and Beverdige, 2003).

Except few, not much information is available about the herbicide application in an orchard of seabuckthorn. A number of chemicals have been used to check the growth of weeds in seabuckthorn plantation (Schroeder and Alspach, 1995). For *Agropyron repens* and other grasses, Dalapon or Sodium trichloroacetate may be used. When rosettes of sow thistles emerge, they are treated with amino salt 2,4 D. Later on *Convolvulus arvensis* may emerge, which is also killed by 2,4 D. We can also use 2 M 4 x M (tropotox) instead of 2,4 D.

In 12-15 days after herbicide application, a treatment with flat cutter or cultivator is carried out for removal of other weeds. We can use 2 M-4 XM (Tropotox), instead of amino salt 2.4 D or interchange them. Removal of perennial weeds is very simple, when using rounds up. In this method, two-lobed weeds as well as grass weeds die. The use of round up allows shortening the period of preliminary soil preparation, if round up has been applied during the period of vigorous growth of weeds. It's more profitable to use tank mixes of round up with amino salt 2.4-D.

In second year, Symazine and Promentryn are used for young weeds. Herbicides are used early spring before weeds emerge or late autumn after leaf fall. In water erosion sites, herbicides may be used during early spring. When *Salsola collina* grows in seabuckthorn plantations, Symazin application is not effective. When annual grasses like *Setaria glauca, S.viridis* and *Echinochloa crusgalli* emerges, it is useful to apply Promentryn, as root and contact action preparation. In plantations with *Capsella bursa pastoris*, medium application of Prometryn is not effective. In this case, it is better to use tank mixture of both herbicides (half dose for each) (Levandovsky, 2003).

In seabuckthorn orchards in Russia, only "Raundap" and "Prometrin" have been permitted. Raundap has a wide spectrum of action and used at the concentration of 4 litre/hectare at the cost of $ US 700 per 100 hectares; which control majority of the weeds. As a precaution, in small plantations, spraying of seabuckthorn leaves from this herbicide is avoided, whereas, it is practically not possible in big plantations. Prometrin is a soil herbicide, which controls the annual weeds. It is used during a short period of favorable weather conditions like damp and warm weather before spouring of seabuckthorn leaves. It is used at the concentration of 4 up to 6 litre/hectare, depending on the soil fertility, with a cost of 1000-1500 US $ for 100 hectares.

The monocotyledonous weed-control is much easier than dicotyledonous weeds because of dicotyledonous nature of seabuckthorn, *e.g.* using of some anti-gramineous herbicides. Preliminary trials in Russia, have passed "Lontrel" at the content of 0.6 litre/hectare, which destroy a majority of dicotyledonous weeds and does not injure a plant of seabuckthorn. Therefore, using a mixture of Lontrel with anti-gramineous herbicides is the most effective chemical control of weeds in seabuckthorn at the cost of about 1500 $ US for 100 hectares. We can apply the mixture during the period from the middle up to the end of June during mass growth of weeds. The scientific studies have found total absence of the chemicals in seabuckthorn fruits up to the stage of harvesting. However, it is yet to be permitted in commercial plantations of seabuckthorn in Russia.

However, German experts have cautioned to avoid high doses of the chemicals/herbicides. After 5-6 years of plantation, when the lower part of seabuckthorn crown is surrounded by root-proliferous perennial weeds, it is possible to carry out treatment with the help of systemic herbicides, such as round-up, without applying the solution drops on the leaves of seabuckthorn cultivars. Ratoon sprouts from horizontal roots may be cut and used elsewhere or destroyed, as they create problems in the management of the stand. Generally they are borne after 3 4 years of plantation.

Chemical tests of seabuckthorn fruits showed that "Symazine" and "Prometryn" were not found in the fruits before their harvesting. Prometryn is absorbed in the plant quicker than symazine, but it was not found even in shoots and leaves, 2 months before the harvesting. Chemical tests also show that a definite regularity exists in occurrence of Symazine residues in the green fruits of seabuckthorn. When in the first half of the period, beginning from herbicide application till harvesting of fruits, when the rainfall is less than the average perennial norm, and in the second half, rainfall exceeds the average perennial norm; the appearance of chemical residues is inevitable. When there is more precipitation deviations in the second half of vegetation from the average perennial norms, there are higher contents of residues in the berries. As they have been found in the ripe fruits, application of such herbicides like flan and dalapon in seabuckthorn plantations is not advisable from the health point of view. It is concluded that judicious use of herbicides along with other physical means help in the eradication of the weeds in the seabuckhorn stands.

Furrows

Digging of 30-40 cm deep furrows at 40-50 cm distance on both sides of seabuckthorn row, helps in alleviation of root competition with crops as well as accumulation of moisture near the root system of seabuckthorn plants in the orchard (Plate 12).

PRUNING

The main objective of pruning of seabuckthorn branches is to facilitate fruit collection and better growth of fruit bearing branches by more exposure to sun light. The specific objectives of pruning of seabuckthorn may be to develop and maintain proper size, shape and architecture of the tree; to improve the branching habit and strength; to cut the old and low fruit bearing branches with poor fruit quality; to produce and maintain the optimum amount of new and young fruit bearing branches; to improve the plant vigour by removing broken, diseased, or pest-infested branches; to improve the light penetration within the tree canopy; to induce and maintain annual fruit yield, to renew the branches with decreasing the fruit bearing and to improve pest control by ensuring effective penetration of light, management and application of pesticides (Albrecht *et al.*, 1984; Beldean and Leahu, 1985). A tree, which is pruned annually, appears to increase the vigour, as it produce the new and better shoots for fruits, where the pruning cuts are made on the branch. Moderate pruning will increase the fruit yield and fruiting life of the plants (Savkin and Mukhamadiev, 1983; Albrecht *et al.*, 1984).

Size of a tree size is known to have a significant effect on the ratio between vegetative and reproductive growth within a tree. Large tree produces less fruit per unit of vegetative growth than small tree. The earlier the regular pruning is initiated; the greater is the control of tree size. Tree size determines the percentage of the total leaf surface that receives the adequate sun light. Reducing the tree size from 5 to 2.5 m reduces the shaded interior from 24.4 per cent of the tree volume to 1.6 per cent. The height of seabuckthorn stems can be maintatined at 1.5 m (Plate 13).

After 4 years of plantation, seabuckthorn plants develop the crown at the height of 2-3 m and forms its crown at the base of the main stem. The crown should be pruned annually to remove overlapping branches and long branches should be cut to induce the development of lateral shoots. The main truck stops gaining height in 5th year and branches begin to grow from the lateral buds. In case the plant canopy is dense, they should be thinned to allow light penetration. Excessive application of nitrogen fertilizer also induces the vigorous vegetative growth. High vigor may initially increase the fruit yield up to a point, but eventually shoot growth becomes directly competitive with fruit set, resulting in a decline of fruit yield. Seabuckthorn exposed to more light has been found to produce more fruits than under partial shade. In order to protect from premature senescence of a seabuckthorn tree, 3-year-old branches should be pruned for restoration of rejuvenation. In Russia and Germany, studies found it useful to carry out pruning in creating hedges for mechanical harvesting of fruits. But this method of pruning for mechanical harvesting is not ideal, as the mechanical harvester also cut fruiting branches, which will not bear fruits next year.

Pruning is started the year trees are planted in the field. Pruning may be done during late winter, before bud breaks, when the trees are dormant. Generally, it is done during early spring. Pruning is preferred prior to bud break in the branches. The benefit of this is that it stimulates the vegetative growth, which encourages the healing in the cuts on the branches. Pruning during the spring is usually not done; otherwise it may cause sap to run down along the pruning cuts, which may promote the development of diseases later in the season.

A number of pruning tools are available in the market. For the small branches up to 1.5 cm in diameter, we can use secateurs. For the more thick and larger branches, one can use the pruning saws or loppers. One needs to arrange the pneumatic pruners for large orchards to save the time and do the work easily. One also needs to maintain and keep the pruning tools clean and sharp for the better results. Bigger pruning machines are also avaibale in the market (Plate 14).

We need to make the pruning cuts on the branches should be flush, smooth and made at the proper angle to avoid the delay in the healing process and reduce the susceptibility of the plant to diseases. It is necessary to carry out the wound dressings on the cuts, except in case of large pruning cuts.

The basic principal of pruning in seabuckthorn is based on the view that pruning should be carried out in the branches to replace them in a specified order, usually over 3 to 4 years. The extent of pruning to be done annually depends on the type of species, stage of plant growth and planting space between the rows and plants etc. Normally, we should not remove more than a third of the top crown during a dormant pruning season, and a lesser amount of pruning, if summer pruning is to be performed.

Different locations or zones within the tree have different pruning requirements; therefore they will respond differently. Crowding of branches is one of the most common problems in seabuckthorn orchards, particularly in its high-density plantations. In order to get the maximum fruit yield, one requires a row space filled with effective leaf surface, but this will lead to some crowding of branches. Excessive pruning may also stimulate highly vigorous, unfruitful growth, which may create more shade within the tree than before. In order to control the over crowding, therefore one needs to avoid the extensive heading-back cuts. Therefore, it is better to make relatively few but large cuts than to make many small cuts. Careful observation of the tree will usually identify one branch, which cause most of the crowding in a given area. The cutting of this branch solves the crowding problem with a minimal disturbance of the remaining bearing surface.

Pruning in Mature Trees

On reaching a desired size, tree height must be checked to ensure a proper distribution of nutrients within the tree. Allowing trees to achieve a tall height may result in lesser production of fruits in the lower part of the tree due to shading effect. Large branches are a problem in a mature tree due to harvesting problem, especially in the upper part. The branch: trunk diameter ratio may need to be decreased to 1:4 (25 per cent). The higher the planting density, the lower the branch: trunk ratio should be allowed. In case, the vegetative growth of seabuckthorn plants gets out of hand, we have to consider the root pruning during early spring. We have to use a sharp blade 30 cm away from the trunk and 30 cm deep in the soil to cut the feeder roots and curtail vegetative growth of seabuckthorn plants. Irrigation schedules may have to be changed, in case the roots are pruned in this manner.

Pruning Systems

There are 2 pruning systems, which can be considered in seabuckthorn orchard:

Modified Central Leader

The vigour of the side branches should be reduced to encourage the development of a central leader. After the plantation in the field, plants are headed at the height of about 60 cm, which results in the formation of several upright and narrow angled branches and the top shoot becomes the central leader during the first season of growth. At the end of the second growth season, all the new branches developing within 30 cm above ground should be removed. All the branches are given headed back cut to maintain the dominance of the central leader. In case of over crowding of branches, thinning cuts may be also be given. To maintain a desirable height of tree at 2.0 m or 2.5 m, the central leader should be cut after 4 years of growth. There are some of the following important parameters for the modified central leader:

1. *50 per cent rule*: For the central leader, we remove the branches larger than 50 per cent of leader diameter.
2. *Remove excess wood*: We remove the excessive and old branch, improve the light exposure and rejuvenate the wood even if it is lesser than 50 per cent of trunk diameter.
3. *50 per cent rule on co-dominant branches*: We remove the side wood larger and thicker than 50 per cent of the branch diameter.
4. *Spurs*: We cut the spurs underneath and cut back extensively the weak and hanging branch.
5. Tree height: We provide cut to the trunk at the height of 1.5-2.5 m, when the tree is in full cropping stage, maintaining a low vigour shoots at the top, while keeping

Plate 12: Digging of 30-40 cm Deep Furrows on Both Sides of Seabuckthorn Row

Plate 13: Cutting of Seabuckthorn Stems at 1.5 m Height

Plate 14: Pruning/Training of Seabuckthorn Plants Before Bud Brust in Winter

the lower part of the tree open and vigorous enough to enhance young and healthy flower buds.

6. We head back the branches, stiffen up and keep tree compact with healthy branches close to the trunk.

Open Centre Pruning

Generally, an open centre system has the characteristics of an upright growth habit. However, in case of seabuckthorn, a shrub in nature, with proper pruning to control its growth, the application of an open centre system may be useful to consider. Young plants of seabuckthorn may be pruned at approximately 60-80 cm height and all the buds below 30 cm are removed. Plants are trained to a multi-leader system with major branches extending up and away from the trunk, giving a vase-shaped tree. At the end of the second year, all the newly developed branches below 30 cm are removed. We remove the potential central leader, head back and thin all the branches to maintain the desired height (not more than 2 m) of the plant. Pruning of seabuckthorn plants has to be carried out annually. The fruits bearing zone of the plant moves higher and to the outside of the canopy and the tree gradually gains an umbrella form. Light penetration into the plant canopy is very important to produce quality fruit. In this system, the tree canopy may produce shaded zones, which eventually may become unfruitful. However, the upper surface of the canopy, to a depth of approximately 90 cm, receives adequate sunlight to produce a lot of fruits of high quality. It is useful therefore to minimize the shaded zones within the plant canopy of seabuckthorn during the pruning.

INTERCROPPING

Seabuckthorn is known to allow the healthy growth of grass and other herbaceous vegetation under its canopies due to nitrogen fixing ability of the plant. Not much information is available regarding interaction between various crops and seabuckthorn, when grown together. Seabuckthorn trees (*H. salicifolia*) are known to improve the growth of buckwheat, however badly affecting the growth of peas and poatoes upto 4 and 2 m respectively distance from tree rows (Singh, 2001). Germany, India and China has developed intercopping system of seabuckthorn with various crops (Plate 15 a, b, c and d) (Plate 16). Seabuckthorn mix forests can also be raised with other forest trees like willow, poplar (Plate 17) and rubinia etc.

Plate 15: Intercopping Systems of Seabuckthorn

Plate 15 (a)

Plate 15 (b)

Plate 15 (c)

Plate 15 (d)

Plate 16: Intercropping of Seabuckthorn (*Source*: Lu Rongsen)

Plate 17: The Mixed forest of Seabuckthorn with Poplar in China (*Source*: Lu Rongsen)

DISEASES AND PESTS

Each region has different pests and diseases, which can attack seabuckthorn plantations, *e.g.* in western Siberia, fusarium fungus and seabuckthorn fly are the extremely dangerous to seabuckthorn plants.

Diseases

Not many diseases are known to affect seabuckthorn. However, among them, the distribution area of *Fusarium sporotrichiella* is the biggest and most serious damaging area. It used to broke out largely in Russia in the 1960s and has spread all over the world recently. In addition, some diseases such as verticillium wilt (*Verticillium albo-atrum), Monilia altaica, Stigmina hippophaes, Alternaria* sp., *Valsa* sp. and *Pythium* sp. have caused some damage in Russia and Liaoning, Shanxi and Gansu provinces in China. The disease, which has broken out on the largest scale and damage most severely is *F. sporotrichiella* in China. Since, it has infested in the large berry cultivars of seabuckthorn, it has become the main obstacle with the development of the large berry cultivars of seabuckthorn in China. Verticillium wilt in seabuckthorn is caused by *Verticillium albo-atrum* and *V. dahliae*. The symptoms of yellowing and progressive wilting of the leaves, the presence of discoloured vascular tissue, also the non-symmetrical development of leaf symptoms are typical. The affected fruits develop colour before ripening and dry upand shrivel. Swelling of red colour appear on the bark and stem carks at the end of summer, resulting in the mortality within 1-2 years. There is no treatment, therefore affected seabuckthorn plant should be uprooted and burnt. During seedling stage of seabuckthorn, soil borne *Alternaria, Fusarium* and *Botrytis* may attack and cause the damping off of seabuckthorn plants.

Attack by fusarium is considered as the most dangerous disease in seabuckthorn plantation, which may also cause the failure of its orchards. Occurrence of fusarium leads to sudden yellowing of leaves, darkening of a bark with the further dying off of the dried branches and plants as a whole.

While no control method, including plantation management techniques is known to check this disease, it is possible to develop and use seabuckthorn varieties resistant to disease, but their productivity may be lesser to the best ones. Further, fusarium prone varieties of seabuckthorn should not be planted, and seabuckthorn should not be planted in an old seabuckthorn field. In fusarium prone plantations mortality upto 50 per cent, even more is expected. However, 10 per cent losses of plants by fusarium withering is normal case. Despite several recommendations, the best solutions seem to have selections of disease resistant forms of seabuckthorn. Lesser humidity conditions have been found to reduce the occurrence of diseases. Although, attack of fusarium wilt is not common, but it should be treated as a serious problem in seabuckthorn. Generally, these diseases are secondary and followed dieback of branches or stems in seabuckthorn caused by physical or chemical damage (Melnik *et al.*, 2005). Scientists at Agriculture Research Centre of Finland, also isolated fungus *Stigmina* Sacc. (*Thyrostroma* Hohnel) from stem canker of seabuckthorn. The optimum growth and spore production was at 15-20ºC. Isolated fungus on the buds of susceptible seabuckthorn in autumn caused small cankers around the buds during autumn and spring. The lesions formed around the buds are first reddish and later turn dark brown and sunken. The cankers are elliptical, measuring from 0.5 to 7 cm in length and girdle the stems, which wilt or produce abnormal growth. *Stigmina* sp. produce spores on dead buds and cankers. It causes stem cankers on stems and twigs of seabuckthorn. The mycelium grows under the bark and protrudes the cracking the surface. Most heavily infected were Russian varieties from Altai region and their progenies. Stem cankers

Plate 18: Attack of Seabuckthorn Carpenter Worm (*Holcocerus hippophaecolus*) Larvae

was affected by climatic conditions, damage was more serious, when autumn temperature was high.

Pests

Seabuckthorn fly is the extremely dangerous to seabuckthorn plants in Russia. In seabuckthorn plantations, insect infestations of aphids, rose leaf roller, gypsy moth, gall tick, comma-shaped scale, thrips, and two-spotted mites have been reported. Green aphids (*Capithophorus hippophae*) have been observed feeding on tender shoot tips of seabuckthorn plants, damaging them seriously. Thrips have been observed in late spring and early summer. In Saskatchewan, Canada, occurrence of uglynest caterpillar (*Archips cerasivorana*), fruittree leafroller (*Archips argyrospila*), speckled green fruitworm (*Orthosia hibisci*) and the rusty tussock moth (*Orgyia antiqua*) have been found on several plants of seabuckthorn. These insects are defoliators and commonly found on most shrubs.

In China and Russia, *Rogoletis batata* Hering., *Psylla hippophaes* Frst., *Lymantria dispar* (Linnaeus) and *Malacosoma neustria testacea* Motschulsky have caused damage to seabuckthorn trees in some degree in China and abroad. In addition, some pests such as *Gelchina hippophaeslla* Schak., *Vasates* sp., *Gapithophorus hippophaes* Walk., *Ahyips rosana* L., *Zeuzera coffeae* Nietner, *Thrips* sp., *Labidura* sp., *Chionaspos salicis* (Linnaeus), *Sitotroga* sp., *Acherontia styx* Westwood, *Brahmina coriacea* Hope, *Plodiainter punctella* (Huber) and *Holotrichia longipennis* (Br) have caused damage to seabuckthorn woodland in other countries.

In recent years, seabuckthorn carpenterworm (*Holcocerus hippophaecolus*) has attacked seabuckthorn populations in Inner Mangolia, Shaanxi, Shanxi, Liaoning, Hebei and Gansu provinces causing heavy economic damage to seabuckthorn industries in China. The pest causes damage to stems and roots of seabuckthorn plants during larvae stage (Plate 18). The larvae moves from stem to the root before winter, making the whole roots hollow and causing the death of plants (Luo *et al.*, 2007).

For control, in case of *H. hippophaecolus*, fumigating with aluminium phosphide has proved to be an economical, effective and feasible one. When the average daily temperature for spring is between 21~25°C, the control efficiency is 82.61 per cent at the dose of 1 pill (about 3.2 g) of aluminium phosphide per. At present, no other chemical is registered for the control of insect in seabuckthorn. Growers may consult the agronomists for the chemical control of diseases and pests. However, organic control measures like dormant oil can be used before bud break to prevent occurrence of some pests. There is no chemical control method has been studied. Seabuckthorn plants infested with pests should be uprooted and burnt to check the spread of pests. Lissonota setosa, a natural enemy, may be used against *Holcocerus hippophaecolus* at larvae stage. Wireworm can also be utilized to check this pest. Sex pheromone also has the potential in controlling the *Holcocerus hippophaecolus*. Experts should also work on the application of natural pesticides and development of pest resistant cultivars of seabuckthorn (Luo *et al.*, 2007).

There are also a number of insects, which may damage seabuckthorn plants. Seabuckthorn fly (*Rhagoletis batava*) is the very harmful insect, penetrating the fruit, feeding the fruit soft part and making it unacceptable to the buyers. Commashaped scale (*Chionaspis salicis* L.) sucks the sap from the bark and cause the mortality of the plant. The gall tick (*Vasates* sp.), cause the swelling or galls, disorting the form of the leaves, therefore damaging the them. The gypsy moth (*Ocneria dispar* L.) chews leaves during summer and makes the plant unproductive. The Caterpillars of the moth (*Gelechia hippophaeella*) penetrate into and eats the fresh buds and cause them damage. In Lahaul-Spiti, India, Commashaped scale (*Chionaspis salicis* L.) is believed to be causing huge damage to seabuckthorn in some sites. This has also been found on willow trees. Dath hawk moth (*Acherontia styx*) at larval stage damage the plant. Defoliating beetles Brahmina cariaca has been found attacking seabuckthorn leaves. *Holotrichia lonipennis* and *Plodia interpunctella* cause damage to plant at early stage of plant growth and fruiting stage, respectively (Bhagat *et al.*, 2001).

Birds, mice, can also attack seabuckthorn plants and pocket gophers, which can cause failure of the crop, therefore need their control.

Chapter 17
Harvesting of Fruit

Seabuckthorn fruit ripe during August in Russia, Germany (Plate 19) and other European countries, whereas it matures during September in Himalayan countries like India and China. Overripe fruit are difficult to harvest, as the skin is easily punctured during the harvesting, affecting the quality of raw material. About 15 per cent of the berries burst during harvest, even by hand picking. However this loss is only the juice and seeds can be used as a valuable product for oil producing market.

In some countries, seabuckthorn is harvested in the winter after the berries are frozen. However, this may cause loss of moisture, some nutrients and flavour. Therefore, globally, the most successful method of mechanical harvesting is shaking the individual branches of the shrub, to dislodge the berries, without any damage to berries, if they are not overripe, causing them to fall into a container placed around the base of the plant. Overripe berries carry a strong musky odour with ransid taste, which can be even felt even in the field. Harvesting of over ripe berries may also cause rupture of the berries. In order to remove this problem, berries must be harvested at the appropriate time, quickly transported to the processing unit and be cooled immediately at low temperature (4-6°C) to avoid growth of micro-organism and spoilage of the berries.

The proper timing of harvest and handling of fruit is of high importance in the management of a seabuckthorn orchard. Value of the crop is directly related to how the fruit is harvested and prepared for storage prior to processing. Harvesting of fruit is the most time consuming operation in seabuckthorn cultivation. Fruit can stay on the plant thought out the winter due to the absence of an abscission layer and a firm attachment to the fruiting branch, which also make it difficult in harvesting the berries. The small fruit size, short pedicel, force required to pull off each fruit, the density of fruit on the branch, and the occurrence of thorns in the plant, are the disadvantages during harvesting. Because of the inherent difficulty of fruit harvesting in seabuckthorn due to presence of small fruit stalk and thorns, mainly two principle types of direct and indirect harvesting technologies have been tried and adopted globally. Gupta and Singh (2003) have reviewed various methods of harvesting of seabuckthorn.

Direct Harvesting

In the direct harvesting, the harvester is in direct contact with the seabuckthorn fruit, while harvesting fruits. However, this type of harvesting has been known to cause severe damage to the plant with reduced yield. It includes hand picking, fruit harvesting under frozen conditions, various types of manual harvesters and robotic harvesters.

Hand Picking

It is a most common, labour intensive and time consuming, direct method of fruit harvesting in which the fruits are hand picked from ground along to the top of the branches. A person can harvest 5 10 kgs fruits in 8 hours. The total labour cost measured for harvesting a seabuckthorn plantation of 4 ha was 58 per cent of the total production cost over 10 years.

Harvesting Under Frozen Conditions

The frozen berries during late winter are shaken with wooden sticks and knocked off the branches and collected in a container. The harvesting is generally done in the morning before sunrise. About 40 120 kgs of fruits can be harvested by this method in 8 hours.

Plate 19: Seabuckthorn Orchard (3 m × 1 m), Ready for Harvest of Fruit. Male row around the female rows.

Cutting of Fruit Laden Branches

In Mongolia, China and Russia, 20–25 kg fruits can be collected per day by cutting of fruit laden branches. In a variation to this method, the fruit laden branches after cutting are stored under the roof till late winter, when the berries get frozen. Frozen berries are then separated from the branches through stick beating. Grain seeding machines are used to separate leaves from the fruits.

However, manual harvesting is proving expensive around the world, making the commercial cultivation of seabuckthorn economically less viable, therefore there is a search globally for mechanical means of fruit harvesting. However, there is no suitable mechanical harvester, which could be used for all types of seabuckthorn varieties and field conditions.

Direct Manual Harvesters

Jaw tooth and brush type fruit cutter: In Russia, a hand-operated jaw tooth and brush type fruit cutter device has been designed and developed, for direct harvesting the of seabuckthorn. It consisted of two jaws with teeth and brushes. The jaws are closed on to a fruiting branch (by spring-loaded handles) and the fruits are rasped off and fall down through a fabric chute into a bag-type container. The device weighs only 0.6 kg, and during trials, a worker could harvest 40-60 kgs of fruits in 8 hours, which has 2-3 times better harvesting rate, as compared to conventional hand picking.

Profiled Teeth Combs

In Russia, the designs of 3 simple improved hand tools for the direct harvesting of seabuckthorn were developed. These tools consisted of combs with profiled teeth, spring- loaded runners or 'scythe-type' runners, which combed off the berries into a hand held container. The typical rate of fruit harvesting was 48 80 kg in 8 hours.

Fruit Comb

A manual yet simple direct fruit harvesting method has been developed in Mongolia. In this method, a steel wire hook (clip or loop) is used to harvest fruit by combing barriers in two directions at once. Through this method, one can harvest 40 100 kgs of fruits in 8 hours.

Manual Cutter

A variant of the manual cutting comb based fruit cutter, with massive conical mandrels has been designed, developed and tested. Through harvesting tests, it was proved, that the developed cutters were suitable, depending upon the appropriate timing of the harvest at the beginning of fruit ripening. Experienced workers using this manual cutter could harvest more than 40 kg of fruits in 8 hours, at Czech University of Agriculture, Technical Faculty, Prague, Czech Republic.

Indirect Harvesting

In case of indirect harvesting, the harvester is not in the direct contact of the fruit itself. On the contrary, harvesting is done through shaking of the plant. Generally indirect methods of fruit harvesting have been found to be more efficient and less destructive compared to direct methods of fruit harvesting in seabuckthorn. Common types of indirect harvesters include chemical harvesting, suction air type stream harvester, harvesting rig, vibration based berry harvester, various types of static and mobile mechanical shaker harvesters.

Chemical Harvesting

Ethephon [(2-chloroethyl) phosphoric acid], an ethylene generator at a concentration of 250 ppm decreased the attachment force between seabuckthorn fruits and the branches by 30 per cent. The treatment was shown to reduce the mechanized harvesting time by half and ensured a complete crop harvest. Preliminary study in Canada, has shown that Ethephon, when applied 10 days before harvest, can reduce berry removal force marginally.

Suction Air Type Stream harvester

A suction air type stream indirect mechanical harvester has been developed to detach seabuckthorn fruit from the bush, during the mechanical harvesting, by using force of air suction. During test runs, it was able to harvest up to 99 per cent of the fruits and the productivity was increased by 4 6 times as compared to manual harvesting.

Harvesting Rig

A test rig was developed to standardize the most effective oscillation frequency, amplitude and shaking period for shaking off the fruits from the branches, detached from bushes of 'Hergo" and "Leikora", cultivars of seabuckthorn in Germany. It was found, that optimum amplitude, frequency and shaking period were 25 mm, 25 Hz, and 2s, respectively.

Vibration Based Berry Harvester

In Germany, experts have developed a vibration-based harvester, which can separate seabuckthorn barriers from cut pieces of small fruit branches, without any freezing or chemical treatment of barriers. Though the fruit quality was found to be highly acceptable, but the applicability of same harvester to different varieties of seabuckthorn was highly variable.

Indirect Mechanical Shaker Harvesters

Shakers, as indirect fruit harvesters, were developed at the beginning of 20th century. They generally shake the trunk or main branches of the large bushes and trees. Trunk shakers are mainly preferred over the branch shakers due to their less damage to the plant trunk as compared to the branches. Shaker harvesters function particularly on the basis of two principles for harvesting of bush fruits. One is based on side ways shaking with long plastic rods, specifically suitable for raspberry harvesting. Other shaker harvester use drums with shaking metallic or plastic fingers for back and forth shaking. They have been found to be suitable for blue berry, currants and grapes harvesting besides seabuckthorn fruits.

Inertial Trunk Shakers

Stain *et al.* (1985) successfully tested a shaker at 25 mm amplitude and 18.5 Hz frequency, for harvesting of seabuckthorn berries. To further improve the harvesting performance, an inertial trunk vibration shaker harvester was tried to harvest the seabuckthorn fruit. For this mechanical harvesting purpose "Moldova" variety was chosen, as it had large fruit, high productivity and the fruit detachment force from the branch decreases with time. This harvester operated at a capacity of 29-33 plants/h, with a productivity of 800 900 kgs of fruits in 8 hours (Stain *et al.*, 1985 in Gupta and Singh, 2003)).

German researchers, using seabuckthorn cultivar "Hergo" harvested the fruits, using a mechanical shaker at two different settings of 25 mm and 18 mm amplitudes and 25 Hz and 30 Hz frequencies, respectively, with satisfactory results. Experts at Agriculture and Agri-Food Canada, Pacific Agri-Food Research Center, Summerland, British Columbia, Canada, developed a highly efficient shaker, which could completely harvest fruits from each mature seabuckthorn tree in just one or two minutes, without causing any damage to the fruit, thereby drastically reducing the harvesting labour cost.

In yet another attempt, to further develop the mechanized harvester, for seabuckthorn fruit harvesting, researchers at Swedish University of Agricultural Sciences, Fjalkestadsvagen, developed and tried two different berry shaker harvesters with varying efficiencies and extent of damage to the plants, because some varieties of seabuckthorn could sustain bark damage without any reduction in next year's crop yield. On the contrary, some other varieties, sensitive to bark damage, suffered severely with heavy crop yield loss or no yield at all in the next cropping season, implying thereby, that there is still a scope of improvement for this shaker harvester in a variety independent manner.

Mobile Mechanical Shaker Harvester

Experts in Mongolia developed a mobile electric motor driven mechanical harvester, which was capable of moving between bushes at a rate of 12 km/hour. Its metal pivots, enclosed in rubber case, vibrated the seabuckthorn plant with an amplitude of 10 14 mm at a frequency of 13.2 18.5 Hz, resulting in harvesting of 92 98 per cent of frozen barriers, with a negligible damage to the bushes. The fruit-harvesting rate was 1600 kgs of fruits in 8 hours, through this harvester. It was electricity driven, which was provided either by petrol driven engine or from a tractor shaft. The whole of harvester was mounted on a lorry, which was equipped with vibrating shields and pivots of 8 cm diameter. These were motor driven and through eccentric gears, the amplitude of vibration could be altered, as per the harvesting requirement for different seabuckthorn cultivars.

Two main problems have generally been encountered with the shaker harvesters by seabuckthorn commercial growers world over. One is of the damage to the bush, which is of very serious nature, since the bark of seabuckthorn bush is easily damaged by harvester or shaker pivots and secondly in the same mechanical shaker harvester, frequency, amplitude and shaking time settings, used for fruit harvesting in one variety can not necessarily work with other fruit varieties of seabuckthorn.

Ideal Mechanical Harvester

To over come some of these potential bottlenecks, improved complete harvesters have been developed and are under testing in different countries including Germany, Russia and Sweden. German and Russian complete harvesters, use a tangential shaking drum, which rotate clockwise to move branches through harvester. Fruit laden cut branches are collected on a platform and are fed to a berry separator machine. The Swedish harvesting machine, developed at Department of Agricultural Engineering, Swedish University of Agricultural Sciences, Uppsala, cuts the fruit laden branches with header and sends them to shaking device. Using this complete harvester, some seabuckthorn varieties with acceptable results, could be harvested, while in other seabuckthorn varieties either the fruit skin remained attached to the branch or fruit got crushed, before harvesting.

In Lisavenko Scientific Research Institute of Horticulture for Siberia, Barnaul, Russia, another type of mobile complete mechanical harvester has been developed. It cuts the seabuckthorn fruit laden branches with the help of built in saw disk and these cut branches are fed into the shaker, which separate the fruits from cut branches. In this type of complete harvester the total fruit yield in 10 test varieties, varied from 4.6 tons/ha to 19.6 tons/ha, whereas the undamaged fruit yield varied from 55 per cent to 73 per cent of the total fruit yield.

One of the most modern harvesting machine for seabuckthorn berries has been developed at Humboldt University, Berlin. The machine is run by 6 labours and harvest fruits with 10 per cent damage. The machine harvest fruit with a vibration intensity of 30 Hz. In another model in Germany, cut brannhes laiden with fruits are thrown into fruit separating amchine (Plate 20).

Productivity of 3-year-old fruit bearing branches of seabuckthorn, once cut by the machine is restored only in 3 years. Under dry conditions, heavy cutting of branches may also cause partial drying of the plants. Further, these machines collect by cutting of the fruit bearing branches only upto 60 per cent of fruit on the plants. It is important to properly train the fruit collector in harvesting of seabuckthorn plantations by mechanical means. Productivity of fruit harvesting is much higher than manual methods. It is also important to compare not only productivity but efficiency of fruit harvesting per unit space and time. In manual harvesting, one can collect about 80 per cent of fruits, significantly higher than mechanical harvesting of 60 per cent fruits on the plant, however, there is a significant damage to plants in the manual way of fruit collecting. Therefore, practically productivity of mechanical harvesting is almost equal to manual harvesting. Mechanical harvesting of seabuckthorn fruit also maintain the quality of the fruit collected with minimum contamination of harmful microbes.

As Canadian experience show, comparing to manual method, the mechanical harvesting method usually eliminates the necessity for maintenance pruning, leaving a hedge that has been uniformly cut. This method results in a 2 years harvesting cycle but increases harvesting productivity to 600 kg/hour in comparison to 7 kg/hour by hand picking, with crop waste of 10 per cent. However, this method is only justified on large plantations of seabuckthorn.

In Department of Biosystems Engineering, Faculty of Agricultural and Food Sciences, University of Manitoba, Canada, a team of scientists developed the seabuckthorn harvestor from a machine used for harvesting of berries from sakatoons, chokecherries and other small berry bushes. For these plants, the machine was used at a frequency of 10 Hz and the amplitude to 15 mm. The machine harvested almost 100 per cent of seabuckthorn berries at a frequency of 25 Hz and amplitude of 32 mm. The machine was developed into a fruit harvester of seabuckthorn. This consists of a long arm attached to a plat form trailer, with a clamp, attached to the very end of the arm. The clamp is placed around the base of the tree trunk and tree is shaken by the harvester. Presently, people are using a tarp to collect the berries and working to improve the collection method.

Leaf Harvester

Department of Biosystems Engineering, Faculty of Agricultural and Food Sciences, University of Manitoba, Canada has also been working to develop a leaf harvester. The leaf harvester has two brushes attached to the long arm of the machine that swirl around

Plate 20: Branches Laiden with Fruits Cut by Hands and Thrown into Fruit Separating Machine (Germany)

the seabuckthorn plant and remove the leaves. It takes about 40 seconds in remove all the leaves from a branch.

Economics of Harvesting

Manual harvesting of seabuckthorn berries is time consuming and expense, therefore affordable only in cheap labour countries. In India, farmers collect the fruit by hitting the plant gently with a stick generally from 4 AM to 10 AM (6 hours)), when the fruit stalk is stiff. Farmers pay about Rs. 90 (2 USD) per person per day, while one person can collect about 15 kg. In some rich countries, farmers have to pay up to US 30 USD to workers, collecting 100-150 kg of fruits of seabuckthorn during 8 hours. Therefore, it requires 3000-3500 man-days for harvesting of seabuckthorn fruit in 100 hectares, with a productivity of 5 tons per hectare. This is equal to about 100 skilled workers, which can complete the work to collect fruit in 30-40 days. However, the fruit harvest may vary from variety to variety. In Russia, Chuyskaya, Panteleevskaya, Inya and Elizaveta have been found most suitable for manual harvesting.

In Canada, the non-availability of labour to harvest could pose a problem. Growers should consider contracting out, experience has shown that 3 persons are required to harvest 5000 kg fruit in one acre, over the 25 days period, giving careful instructions to ensure a clean, top quality product, for example, to the grocery shelf. If farmgate product is used for the processing plant then some leaf contamination may be acceptable. Picking shoulder harness trays may be utilized or stripping to ground sheet collection. Note that minimal damage to shrub is experienced by this method, ensuring a harvest the following year.

Chapter 18
Post Harvest Handling

In order to fetch a better price to growers, it is necessary to timely transport seabuckthorn fruit in proper containers and to store the produce at low temeparture before processing units (Beveidge, 2003).

TRANSPORT

Seabuckthorn fruit is soft skinned and perishable; therefore should minimize the transport time of seabuckthorn fruit from the field to the processing unit. Specially designed trays have been found useful in seabuckthorn fruit storage and transport. Fruit should be transported to processing unit under low temperature conditions, in case it takes more than 8 hours in transporting the fruits.

CLEANING OF FRUIT

Cleaning of seabuckthorn fruit is particularly important for removal of dust particles, microbes and undesirable plant parts etc. Cleaning operation includes the removal of damaged fruit and stems, leaves or other debris collected during fruit harvest under the field conditions. The accumulated debris can affect the quality of raw material by creating the unique flavours from spoiled berries or 'green' flavour from leaves or branches. Therefore, washing seabuckthorn fruit with little hot water before the processing may provide a method of reducing this odour and improve the quality of seabuckthorn raw material.

COLD STORAGE

After harvesting the fruit, it should be spread in a layer not more than 15 cm thick in shaded area. After fruit collection from the field, it should be cooled to low temperature as early as possible (within 1 or 2 hours). In order to check the contamination and fermentation, seabuckthorn fruit should be transported to processing unit as early as possible. Fruit should be cooled immediately at 4-6°C or even less to minimize the multiplication of microorganism. Storage place as well as containers (perforated trays/baskets) should be cool air circulated, to allow maximum free cool air circulation of fruit. For storage of more than one day, fruits should be stored frozen by quick freezing technique.

Freezing in shallow trays increases the freezing rate, which helps to maintain the quality of the fruit. Specially designed cooling machine has been developed for other small fruits, which can also be used for seabuckthorn as well. Package the fruit in 2.5 mil plastic bags or plastic lined boxes and store at -18°C to -22°C until processing. Fruit should be stored in sealed plastic bags. However, simple way is to extract the seeds and store the pulp and seeds separately. In some cases the pulp can be dried at freezing temperature and stored as a powder.

In case, the fruit is to be supplied to processing plant at distant place, growers should build a pre-cooler on their property, according to their requirement of fruit production. A simple storage structure with low temperature conditions can be constructed as relatively inexpensive means of fruit storage. Under these conditions, fruit can be stored for a maximum storage in a cooler for 10 days before shipment in a low temperature transport vehicle to the processing plant. After reaching to the processing unit, fruit should be contained in plastic wrapping, and then placed in cold storage at a temperature below freezing (-1 or -2 °C), if processed within 30 days of reaching. Fruit should be frozen at very low temperature (-18 °C) for long-term storage (1 year) without further loss of ingredients. Later, fruit may be thawed and processed to pulp etc. Seabuckthorn pulp or juice extracted by pressing or centrifugal method should be stored refrigerated, pasteurized and frozen for long time storage. Or, they may be processed into pasteurized or sterilized finished products, which can be stored at room temperature conditions. Self-life of finished products is also limited and can be increased by storage at low temperature.

QUALITY PARAMETERS

Quality of seabuckthorn fruit before processing affects the quality of products produced after processing. Seabuckthorn fruit should be collected at the proper time of ripeness, as overripe fruit is deficient in nutrients and develop smell and poor taste, which are also observed in the juice and other products. Occurrence of dust particles, leaves and branches also badly affect the quality of seabuckthorn produces. It is advisable that seabuckthorn fruit should be fresh, free from any contamination, maintained at low temperature (-2-4°C) or stored at low very temperature (-18°C).

There are different opinions regarding time between collection of seabuckthorn fruit and processing, however, there only one opinion that fruit should be fresh and free from contamination and fruit should be processed as early as possible after harvesting from the field.

Skin of the fruit should be firm and should not rupture during transport and storage. Containers, transport and storage conditions should suit seabuckthorn fruit. Maintaining low temperature during transport and storage is important.

Growers or producers of seabuckthorn fruit should agree to the quality assurances of industries. Russian experts recommend that (State standards acting in the Russian Federation and CIS), content of branches and other parts of the plant should be lesser than 1 per cent of the total fruit and mineral admixture lesser than 0.5 per cent, the content of fruits, damaged by pests, lesser than 2 per cent and content of non-ripen fruits not more than 1 per cent. Fruit quality (colour, taste, smell and oil composition) also depends on the species, subspecies, genotype and variety. The industry should ensure that the product has always the same good quality of fruit. One solution is that variety of raw materials with varying qualities should be mixed if possible during processing. Another way is the separate storage of the liquid half products of different qualities and their blending for the production of the final products such as nectars and drinks.

PROCESSING

Seabuckthorn fruit can be utilized in the preparation of beer, syrup, jam, squash, juice and ready to serve beverages besides its use of fruit oil in pharmaceutical industry to prepare médicinal and cosmetic preparations. German, Canadian, Russians, Chinese and Indian experts etc. have developed processing technologies for seabuckthorn juice and oil extraction and development of variety of health products. The information regarding processing technologies has been given below in brief and it is just valuable for the sake of knowledge of growers.

Washing of Fruit

There is always a need to inspect the fruit for contamination before the processing even, the suppliers have met the quality assurances. Leaves and twigs should be manually removed from the fruits manually, if they do not meet the quality standard. Seabuckthorn are washed in water to separate the microbes and dust particles attached on the surface of the fruit, as they often come from a dry and sandy places. For washing, emersion of fruit in water is avoided, as there is a risk of dilution of soluble solids in fruits. Generally, for washing one can use a blast (blowing) washer, equipped with indirect or direct heating of the wash water for thawing the frozen berries.

Crushing of Fruit

The juice extraction is carried out by pressing of the fresh or enzymatic fermented berries mash. A serpentine belt press in which the fruits are placed between two belts in a thin layer and then pressed between rollers work better than rack or cloth types of press. Roller mills and hammer chopper/mills are most suitable for crushing the seabuckthorn fruit into a mash. Mills with attachments outside of the body for injecting steam to clean and sterilize are useful. The seabuckthorn berry mash can be further processed into juice or puree. The pulp oil extraction happens from the juice or puree.

There are different suitable presses like hydraulic basket press and belt press (Plate 21). Screw presses are also suitable, but not so effective (lower juice yield, high content of lees parts). Belt presses function continuously. The berry mash is sent from mill into the press and is distributed as consistent shift between two endless belts. The extraction is rapid. The raw juice should pass a sieve for separation of mash parts, which occasionally fall from the belt in the juice. The juice produced by the belt press has a low solid content. The juice yield in seabuckthorn berries is 70-75 per cent, depending on the fruit. Hydraulic or pneumatic presses for fruit processing function intermittently. Great basket presses are built horizontal. Simple vertical hydraulic presses (with basket) might be useful for small processing capacities, therefore not for industrial scale. The press power (3 bar) is produced by water pressure, *i.e.* by connection on a tap water pipe (Gimmler and Waehling 2007).

Seabuckthorn juice obtained by pressing is a complex mixture of large and small suspended soil particles and droplets, contaning about 2-3 per cent oil. The orange colour of juice and oil droplets is due to presence of carotenoids.

Puree

The processing of seabuckthorn fruit into puree is useful as more valuable parts of the berries, like lees with flavonoids, vitamins, oil and pectin are released in the final product, having a higher nutritional value and the yield of the fruit is higher than by normally pressing of the mash. There is higher microbiological safety and lesser oxidation of fruit nutrients due to closed system. There is a more acceptability of cloudy and natural products like nectars and drinks by the consumers. The seeds should be separated from the mash before the heating for a good seed oil quality. The transformation of the mash into puree happens in two or three stages by sieving in two phases (Gimmler and Waehling 2007).

The primary job of the first sieving step is the separation of the seeds from the mash before the mash is heated for the further sieving. The heating should be fast done at a temperature of 80-85 °C, immediately after the first sieving of the fruit. This helps in softening of the fruit parts by partially thermo disintegration, therefore better puree yield; inactivation of fruit own enzymes, specially pectin esterase and no pectin disintegration during the following further processing. Protopectin is released and dissolved, which is advantageous for good lees stability in the puree. Microbes are also inactivated.

Application of double action centrifugal extractor produces the super fine puree extraction. The basic principal is the centrifuging and secondly the squeezing. The produced puree is very fine and contains only a little part of brown seed membranes.

Pulp Oil Extraction

Traditional and old method of oil extraction from seabuckthorn pulp is through steam flow after sedimentation of big lees parts and the watery phases or the extraction of pulp oil by other cooking oils (*e.g.* sunflower oil) or other solvents. The extraction rate of pulp oil is low. The content of the pulp oil is difficult to estimate in such oil mixtures. Generally, the content of beta-carotene is used alone as standard.

During the last one-decade, some simple methods were developed for the (partial) pulp oil extraction, some of them with good acceptable quality. These methods had the main objectives of the oil extraction for high market value in cosmetic and drug industries, often without consideration of a good juice/puree quality, with little market demand initially.

However, recently, demand for seabuckthorn puree/juice has also raised, therefore, a need for standardization of oil extraction methods for optimisation of quality puree and juice. The extraction of the pulp oil has been found advantageous by centrifugation of the puree or juice. This stabilised oil is partially extracted by centrifugation as yellow-orange paste "Seabuckthorn butter", often in combination with small quantities of free pulp oil. The "butter" contains about 30-40 per cent pulp oil and 60-70 per cent juice (Fischer and Albrecht, 2003).

The juice/puree bears a reduced oil content after the centrifugation, but it is not oil-free. Further centrifugation of the "butter" with 3-phases-decanter produces good results (Plate 22). After treatment with suitable enzyme compounds, the oil is easy to separate with centrifugal force, also with 3-phases-decanter (Gimmler and Waehling 2007). Unlike other methods, like the intensive thermal treatment or the destabilization of the oil by an enzymatic treatment of the purée, this method delivers a lower pulp oil yield. However, the method has the following advantages: The puree is not treated with enzymes, no pectin losses: high lees stability, No high temperatures (80-85°C) additional to the mash heating with the result of better retention of vitamins and better taste. The produced beverages from this puree contain still pulp oil and lipophilic vitamins and are more valuable for the nutrition and better taste than oil free drinks. An integrated oil processing technology has been standardized for high pulp oil extraction in India at Regional Research Laboratory (CSIR), Thiruvananthapuram (Ranjith *et al.*, 2007). The oil produced is quite rich in vitamin E and beta-carotene etc. and has high commercial values.

Seed Oil Extraction

The solvent method (hexane etc.) for oil extraction from seabuckthorn seeds has been initially used in Russia and China, but not done any more due to quality and residues problems.

One advanced method has been developed in Germany. It is based on the hydrolyse of milled seeds by addition of water and different enzymes. The oil is separated by centrifugation after the enzymatic fermentation and special treatments of the slurry. One advantage of the process is the higher yield in phospholipids and phytosteroles in the oil, but further research is required to optimise the process (Mothes Hochdrucktechnik GmbH).

The seeds of seabuckthorn contain about 10 per cent oil. The classic extraction processing is the out-pressing of the dried seeds by screw presses, so called screw oil expeller. The temperature of the oil should not be over 40°C for the quality "cold pressed" oil.

Modern oil expeller works in the suitable temperature range. It is useful to pre-crash the seeds before the pressing. The seed oil yield with oil expeller is approximately 40 to 60 per cent of the oil content of the seeds. The residue from the oil extraction, the press cake or oil cake has rest oil content of about 5-6 per cent. The press cakes still contain oil among other substances (*e.g.* cellulose, protein) and is very good suitable for animal fodder. In other case, the rest oil from press residue can be extracted with solvents or CO_2, if it is economical. It is also possible to extract the oil with an oil expeller from dried pomace (skins and seeds from juice extraction with presses or from the first sieving step of the puree line) or from dried berries.

Another method for extraction from seeds is with super-critical CO_2, which produce high oil yield. The CO_2, which is very similar to a liquid under high pressure, takes on the function of the extraction

Plate 21: Fruit Crushing Machine for Separation of Seabuckthorn Pulp and Seeds/Press Residue

Plate 22: Centrifugal Machine for Separation of Oil from Seabuckthorn Fruit Pulp

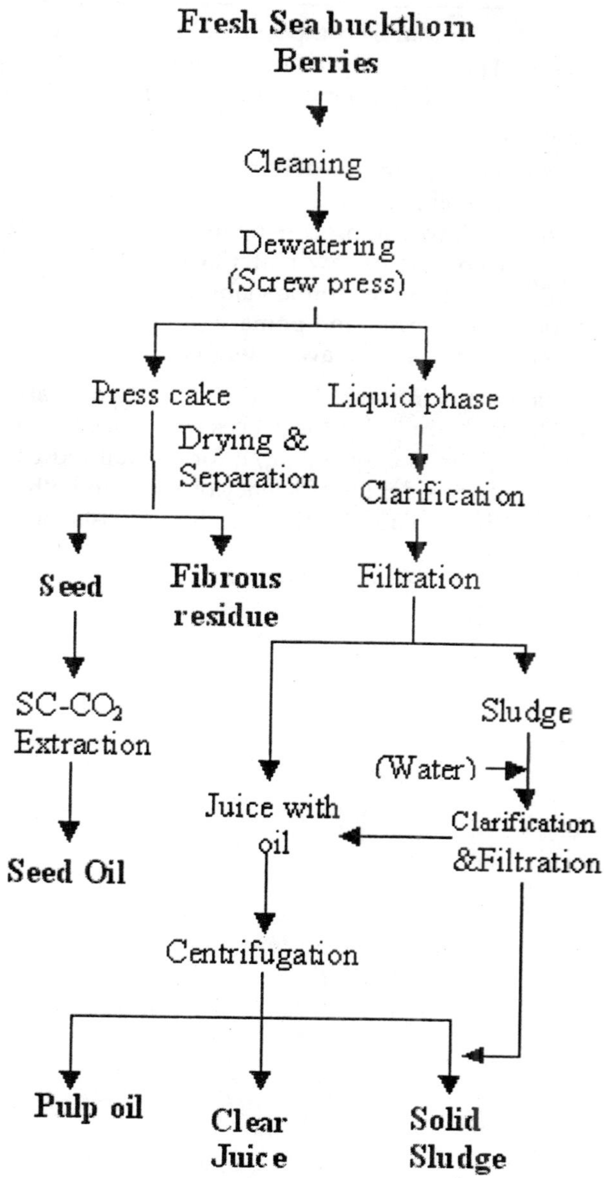

Figure 5: Flow Diagram for the Integrated Processing of Fresh Seabuckthorn Berries (*Source*: Rangit *et al.*, 2007)

solvent, extracts the soluble compound (s) from the product in an extraction step. The working pressure is between 50 and 350 bar for the "CO_2-extraction", the temperature mostly 40-50°C (Gimmler and Waehling 2007).

The method is useful as it is done at low temperatures, exclusion of air oxygen, no solvent residues, extraction compounds free of micro organism. However, processing unit is expensive and there should be high volume of raw material and it should function throughout the year for economic viability. Seabuckthorn seeds, dried seabuckthorn berries and pomace (residues from the juice/ puree extraction) are used as raw materials.

There is always interest about the capacity of seabuckthorn processing units. Technically, the processing capacity should be a minimum of approximately 500 kg/h for a small industrial unit. Then, we have to consider the efficiency of a unit and other factors, like quantity of the available berries and storage costs for the frozen berries.

Literature

1. Albrecht, H.J. 1990. Development of seabuckthorn varieties. *Gartenbau* 37 (7): 207-208.
2. Albrecht, H. J., Gerber, J., Koch, H. J. and Wolf, D. 1984. Experience in growing seabuckthorn. *Gartenhau* 31: 242-4.
3. Antonelli, M., Raffo, A. and Paoletti, F. 2006. Biochemical changes during ripening of seabuckthorn (*Hippophae rhamnoides*) fruits. In: *Seabuckthorn-A Multipurpose Wonder Plant*. Vol. II. N.Delhi, p. 285-312.
4. Avanzato, D., Magherini, R. and Lodoli, E. 1987. Studies on the germination potential of seeds and the rooting ability of cuttings of *Hippophae rhamnoides* L. *Atti Convegno sulla Coltivazione delle Piante Officinali, Trento* 9 10 *ottobre* 1986 p: 411-419. (In Ittalian and English)
5. Avdeev, V.I. 1976. Propagation of *Hippophae rhamnoides* by softwood cuttings under mist. *Kratkie tezisy Dokl 2-i Vses Konf Molodykh Uchenykh po Sadovodstvu* p: 72-74. (In Russian)
6. Bhagat, R.M., Kahsyap, N.P. and Singh, V. 2006. Insect-pests associated with seabuckthorn (*Hippophae rhamnoides*). *Pest Management and Economic Zoology*. (In press)

7. Beldean, E. C. and Leahu, I. 1985. *Hippophae rhamnoides*-A valuable fruit producing pioneer species. *Forstarchiv* 56: 249-53.

8. Beveridge, T.H.J. 2003. Post harvest handling and storage. In: *Seabuckthorn (Hippophae rhamnoides L.): Production and Utilization* (T.S.C.Li and T.H.J. Beveridge, Authors), p. 69-77. NRC Research Press, Ottawa.

9. Bruvelis, A. 1991. Propagation of seabuckthorn (*Hippophae rhamnoides*) by hardwood cuttings. *Verksamhetsberaettelse Balsgaard* (Sweden), p: 194-199.

10. Danilin, M.A. and Grebennik, A.V. 1977. Grafting on (dioecious) trees considering the sex of *Populus tremula, Populus balsamifera* and *Hippophae rhamnoides. Lesnoe Khozyaistvo* 2: 56-57. (In Russian)

11. Dolgacheva, V.S. and Aksenova, N.A. 2003. Propagation, plantation and management of seabuckthorn (*Hippophae rhamnoides* L.). In: Seabuckthorn-*A Multipurpose Wonder Plant-Botany. Vol. I. Botany, Harvesting and Processing Technologies* (V.Singh, Ed.-in-Ch..) Indus Publishing Company, New Delhi, pp. 360-364.

12. Dwivedi, S. K. Singh, R. and Ahmed, Z. 2006. Morpho-biochemical characteristics of seabuckthorn (*Hippophae. rhamnoides*) growing in cold arid Ladakh Himalayas. In: Seabuckthorn-*A Multipurpose Wonder Plant-Botany. Vol. II.* (V. Singh, Ed.-in-Ch.), Daya Publishing House, New Delhi, pp. 151-158.

13. Eliseev, I.P. 1976. Biologically active substances in the fruits of *Hippophae rhamnoides* in Cnetral Asia and the Caucasus. *Biol aktiv veshchestva plodov I yagod* p. 161-163.

14. Eliseev, I.P. and Mishulina, I.A. 1972. The effect of the minor elements on the increase in seabuckthorn (*Hippophae rhamnoides*) seed germination. *Trudy Gor 'kovskogo Sel'skokhozyaistvennogo Instituta* 38: 110-111.

15. Fefelov, V.A. and Eliseev, I.P. 1986. Biology of germination and emergence of seabuckthorn seeds of different ecological and geographical origin. *Biologiya, khimiya, introduktsiya I selektsiya*

oblepikhi p: 110-115. (In Russian)

16. Gakov, M.A. 1980. Prospects for the development of *Hippophae rhamnoides* in the Tuva, USSR. *Lesn. Khosz.* 2: 51-52.

17. Fischer, M. and Albrecht, H. J. 2003. The Seabuckthorn collection in Germany-An Example for the Necessity of Keeping and Sustainable using of Genetic Resources. In: *Proceedings on the 1th Seabuckthorn Congress of the ISA*, Berlin.

18. Garanovich, I.M., Shapira, D.K. and Naryzhnaya, T.I. 1983. Mineral fertilizers and fruit quality of seabuckthorn. *Vestsi Akademii Navuk BSSR, Sel 'Sagaspadarchykh-Navuk.* 4: 79-81.

19. Gardner, I.C., Clelland, D.M. and Scott, A. 1984. Mycorrhizal improvement in non-leguminous nitrogen fixing associations with particular reference to *Hippophae rhamnoides* L. *Plant and Soil* 78: 189-199.

20. Garonovich, I.M. 2003. Introduction of *Hippophae* L. in Belarus. In: *Seabuckthorn-A Multipurpose Wonder Plant. Vol. I. Botany, Harvesting and Processing Technologies* (V. Singh, Ed.-in-Chief). Indus Pub. Co., N.Delhi, p.137-151.

21. Gatner, E.M.S. and Gardner, I.C. 1970. Observation on the fine structure of the root nodule endophyte of *Hippophae rhamnoides* L. *Arch. Microbiol.* 70 (3). 183-196.

22. Geetha, S., Basu, M., Jayamurthy, A. S., Malhotra, A.S., Pal, K., Prasad, R., Kumar, R. and Sawhney, R.C. 2007. Protective and therapeutic potentials of seabuckthorn (*Hippophae rhamnoides* L.). In: *Seabuckthorn-A Multipurpose Wonder Plant. Vol. III. Advances in Research and Development* (V.Singh, Ed.-in-Chief). Daya Pub. Co., N.Delhi. (In press)

23. Geetha, S., Ram, M. S. Singh, V. Ilavazhagan, G. and Sawhney, R.C. 2002. Anti-oxidant and immunomodulatory properties of seabuckthorn (*Hippophae rhamnoides* L.)-An in vitro study. *Journal of Ethnophamacology* 79: 373-378.

24. Gimmler, G. A. and Waehling, A. 2007. Processing of Seabuckthorn (*Hippophae rhamnoides* L.)-An Overview. In: *Seabuckthorn-A Multipurpose Wonder Plant*. Vol. III. *Advances in Research and development* (Singh, V., Ed. in Chief). Daya Publ., New Delhi. (In press)

25. Golubinskaya, N.S. 1972. Some data on raising sea buckthorn seedlings using fertilizers and fungicide TMTD. *Voprosy Lesopark, Khozyaistva I Ozeleneniya Novosibirskogo Nauchnogo Tsentra.* p: 152-160.

26. Gonchar, M.T and Saban, B.A. 1986. Choice of tree and shrub species for afforestation of sites after opencast mining of sulphur. *Lesnoe-Khozyaistvo* 9: 47-49.

27. Gulati, A., Maitra, A., Leelaveni, A., Bhardawaj, S.K., Prasad, R. and Singh, V. 2001. Mycorrhizal associations in seabuckthorn (*Hippophae rhamnoides L.*) from Lahaul and Spiti Valley in Himachal Pradesh, India. In: *Seabuckthorn-A Resource for Health and Environment in Twenty First Century* (V.Singh and P.K.Khosla, Eds.). Souvenir of International Workshop on Seabuckthorn, February 18-21, New Delhi. (Abstract)

28. Gupta, R.K. and Singh, V. 2003. Harvesting technologies of seabuckthorn fruits. In: *A Multipurpose Wonder Plant.* Vol. I. *Botany, Harvesting and Processing Technologies* (V. Singh, Ed. in Chief). Indus Publishing Company, New Delhi, pp. 435-442.

29. Heilscher, K. and Lerber, S. 1996. Process for cold recovery of clear juice, solids and oil from seabuckthorn berries and use of these products. *Patent number DE 44 31 394 C1* (DE4431394C1), Germany.

30. Jike, Z. and Xiaoming, Z. 1992. Progress of study on *Frankia* in nodules of seabuckthorn. *Hippophae* 2: 4-10.

31. Kalinina, I. 1988. Seabuckthorn. *Science in the USSR.* March-April, pp. 80-87.

32. Kallio, H., Yang, B, Tahvonen, R. and Hakala, M. 1999. Composition of seabuckthorn berries of various origins. In: *Proc Int. Workshop on Seabuckthorn,* August-September, Beijing, China, pp. 13-19.

33. Kallio, H., Yang, B. and Peippo, P. 2002. Effects of different origins and harvesting time on vitamin C, tocopherols, and tocotrienols in seabuckthorn (*Hippophae rhamnoides*) berries. *J. Agric. Food Chem.* 50: 6136-6142.

34. Khabarov, S.N. 2003. Elements of commercial cultivation of seabuckthorn in Siberia, Russia. In: Seabuckthorn-*A Multipurpose Wonder Plant-Botany.* Vol. I. *Botany, Harvesting and*

Processing Technologies (V.Singh, Ed. in Chief). Indus Publishing Company, New Delhi, pp. 347-351.

35. Kondrashov, V.T. 1994. New technology of creating stands of seabuckjthorn and other horticultural crops. *Russian Agricultural Sciences* 11: 24-28.

36. Kondrashov, V.T. and Kuimov, V.N. 1987. Vegetative propagation of *Hippophae rhamnoides*. *Sadovodstvo* 6: 13-16. (In Russian)

37. Korzinnikov, Yu.S., Glazunova, E.M., Yusufbekov, Kh.Yu., Krymskaya, N.B. and Gachechiladze, N.D. 1981. Comparative characteristics of seabuckthorn in the western Pamirs. *Otd. Biol. Nauk* 4: 56-61. (In Russian)

38. Korzinnikov, Yu.S., Glazunova, E.M., Yusufbekov, Kh.Yu., Krymskaya, N.B. and Gachechiladze, N.D. 1981. Comparative characteristics of seabuckthorn in the western Pamirs. *Otd. Biol. Nauk* 4: 56-61. (In Russian)

39. Kuznetsov, P.A. 1995. Effect of pre-planting treatment and plastic mulch on rooting of seabuckthorn hardwood cuttings and transplant quality. *Biol. Aspek. Intro. Selek. Agrotekhn. Oblep.*: 159-163 (From *Hort. Abst*. 56: 8687).

40. Levandovsky, V.N. 1993. Fighting with the weeding vegetation in the plantations and nurseries of seabuckthorn. In: *Proceeding of International Symposium on Seabuckthorn*. Novosibink, Russia, pp. 320-321.

41. Li, T.S.C. and Beveridge, T.H.J. 2003. *Seabuckthorn (Hippophae rhamnoides L.): Production and Utilization*. NRC Research Press, Ottawa.133p.

42. Li, T.S.C. and Wardle, D.A. 1999. Effects of seed treatments and planting depth on omergence of seabuckthorn species. *HortTechnology* 9: 213-216.

43. Li, T.S.C. and Schroeder, W.R. 1996. Seabuckthorn: A multipurpose plant. *HortTechnology* 6: 370-380.

44. Lu Rongsen 1990. Research on seabuckthorn (*Hippophae* L.) resources in China. *Acta Horticultural Sinica* 17:177-183.

45. Lu Rongsen. 1992. *Seabuckthorn-A Multipurpose Plant Species for*

Fragile Mountains. ICIMOD Occasional Paper No.20, Kathmandu, Nepal, 62 p.

46. Lu Rongsen 2006. Biochemical characteristics of seabuckthorn (*Hippophae* L.). In: Singh, V. (Ed. in Chief). 2005. Seabuckthorn- *A Multipurpose Wonder Plant- Botany*. Vol. II. *Biochemistry and Pharamacology*, Daya Publ., New Delhi, pp. 98-107.

47. Luo, Y., Zong, S. and Xu, Z. 2007. Integrated management of the main diseases and pests of seabuckthorn. In: *Seabuckthorn- A Multipurpose Wonder Plant-Vol. III. Adavances in Research and Development* (V. Singh, Ed.-in-Chief), Daya Pub. Cop., N.Delhi.

48. Ma, M., Cui, C. and Feng, G. 1989. Studies on the fruit character and biochemical compositions of some forms within Chinese seabuckthorn (*H. rhamnoides*) in Shanxi, China. In: *Proc. International Sym. Seabuckthorn*, Xian, pp. 106-113.

49. Martemyanov, P.B. and Khromova, T.V. 1985. Agrotechnical measures for advancing woody plant growth. *Byulleten' Glavnogo Botanicheskogo Sada* 138: 45-48 (*Hort. Abstr*. 56: 8066).

50. Melnik, V., Borisov, B., Qinxiao, W. and Daiqiong, L. 2005. Diseases and pests of seabuckthorn in the Loess Plateau of northern Shaanxi. In: *Abstracts of 2nd International Seabuckthorn Association's Conference*, Aug. 26-29, 2005, Beijing, pp. 22-23.

51. Min, L., Shuchen, W. and Li, Z. 1989. The role of *Hippophae* in soil and water conservation on Loes Plateau. In: *Proc. Inter. Sym. Seabuckthorn*, Oct., 1989, Xian, China, p. 263-268.

52. Mishulina, I.A. 1976. The effect of foliar nutrition with minor elements of *Hippophae rhamnoides* on the fruit content of biologically active substances. In: *Biol. Aktiv. Veshchestva Plodov I yagod: Buglova, T.*: Frost damage of flower primordial, flowering and cropping of *Hippophae rhamnoides* p: 97-99.

53. *Mothes Hochdrucktechnik GmbH*, D-12489 Berlin, Germany

54. Osipov, Y.V. 1983. Propagation of seabuckthorn by cuttings. *Sadovodstvo* 12: 20-21. (In Russian)

55. Pletneva, T.M. 1983. Intensification of seabuckthorn nurseries. *Agrotekhnika i Selektsiya Sadovykh Kul'tur, p: 149-154.* (In Russian)

56. Poluparnev, Yu. I. and Dulgov, N.P. 1988. A seed separating machine for small juicy fruits. *Lesnoe-Khozyaistvo* 1: 50-51.

57. Potapov, F.F. 1978. New methods of seabuckthorn propagation from green cuttings. *Rastitel'nye Resursy* 14 (2): 215-220. (In Russian)

58. Predeina, R. V. 1987. Fertilization of *Hippophae rhamnoides* in the Altai region. *Sadovodstvo* 6: 13-18 (*Hort. Abstr.* 58: 2046.)

59. Privalov, G.F., Solonenko, L.P. and Skuridin, G.M. 2003. Mutagenesis breeding of seabuckthorn (*Hippophae rhamnoides* L.). In: *Seabuckthorn-A Multipurp[ose Wonder Plant.* Vol. I. N.Delhi, pp.194-210.

60. Quan, H. 2003. Breeding of seabuckthorn (*Hippophae rhamnoides* L.) in China. In: *Seabuckthorn-A Multipurp[ose Wonder Plant.* Vol. I., N.Delhi, pp.187-193.

61. Quanzhong, L., L. Peiyun, C. Shaozhou, B. Yuzhen, P. Lixin and K. Qingjie 1989. Multiple effects of artificial common seabuckthorn (*Hippophae rhamnoides*) forests in western Liaoning. *Proc. Int. Workshop sym. Seabuckthorn*, p. 288-297, Xian, China. p. 415.

62. Quanzhong, L., Peiyun, L., Shaozhou, C., Yuzhen, B. Lixin, P. and Qingjie, K. 1989. Multiple effects of artificial common seabuckthorn (*Hippophae rhamnoides*) forests in western Liaoning. In: *Proc. Int. Workshop sym. Seabuckthorn*, p. 288-297, Xian, China, p. 415.

63. Ranjith, A., Kumar, K. S., Venugoplan, V. V. and Arumughan, C. 2007. Integrated processing technology for seabuckthorn berries and chemical evaluation of the products. In: *Seabuckthorn-A Multipurpose Wonder Plant.* Vol. III. *Advances in Research and development* (Singh, V., Ed. in Chief). Daya Publ., New Delhi. (In press)

64. Rousi, A. 1971. The genus *Hippophae* L.-A taxonomic study. *Ann. Bot. Fenn.* 8: 177-227.

65. Rousi, A. and Aulin, H. 1977. Ascorbic acid content in relation to ripeness in fruits of six *Hippophae rhamnoides* clones from Pyhäranta, SW Finland. *Ann. Agric. Fenn.* 16: 80-87.

66. Savkin, V. A. and Mukhamadiev, S. M. 1983. Effect of branch detachment from seabuckthorn bushes on their regeneration and fruiting. *Ekol. Rast. Zhivot. Zapovedn. Uzhek.* p. 29-32. (*Hortic. Abst.* 54: 8055).

67. Schroeder, W.R. and Alspach, L.A. 1995. Herbicide program at the PFRA sherterbelt Centre. In: *Proc. Western Forest Cons.* Nursery Assn. Kearney, Nebraska.

68. Schroeder, W.R. and Yao, Y. 1995. *Seabuckthorn: A Promising Multipurpose Crop for Saslatchewan.* Prairie Farm Rehabilitation Adminsitration, Agriculture and Agri-Food Canada.

69. Shirko, T. S. and Radyuk, A. F. 1989. Fruit quality of seabuckthorn varieties cultivated in Belorussia. *Vestsi Akad. Navuk BSSR, Ser. Sel'skagaspad. Navuk* 4: 74-79. (In Russian)

70. Shishkina, E.E., Loskutova, G.A. and Arkhipova, T.N. 1985. Biochemical characteristic of seabuckthorn cultivars selected in the Siberian Institute of Horticulture. In: *Biol. Aspects of Seabuckthorn Introduction, Breeding and Agrotechniques*, p. 117-122, Gorky Agricultural Institute, Gorky. (In Russian)

71. Shlyapnikova, A.S. 1985. Efficiency of herbicide application in planting rooted green cuttings of blackcurrants and common sea buckthorn. *Izvestiya Timiryazevskoi –Sel 'skokhozyaistvennoi Akademii* 5: 125-129.

72. Shuhua, H., Xin, Z. and Zhixiang, Z. 1995. Studies on the cutting propagation technique of seabuckthorn. In: *Worldwide Research and Development of Seabuckthorn.* Proceedings of International Workshop on Seabuckthorn, Beijing, China, p. 118-132.

73. Shuhua, H., Zhixiang, Z. and Xin, Z. 1989. Studies on the culture technique of hardwood cuttings of seabuckthorn. In: *Proceedings of International Symposium on Seabuckthorn.* Xian, China. p. 217-229.

74. Siabough, P.E. 1974. *Hippophae rhamnoides*-Common seabuckthorn. In: *Seeds of Woody Plants in the United States.* C.S.Chopmeyer Tech. Co-Ord.USDA-FS Agric.Hdbk. No. 450: 446-447.

75. Singh, V. 1994. A report on arboreal seabuckthorn (*Hippophae salicifolia*) from dry temperate Himalayas. *Journal of Tree Sciences* 3 (1): 67-68.

76. Singh, V. 1995. Rooting rates of hardwood cuttings of seabuckthorn (*Hippophae rhamnoides*). *J. Tree Sci.* 14 (2): 87-88.

77. Singh, V. 2001. Traditional agroforestry practices of seabuckthorn (*H. salicifolia* D. Don.). In: *Seabuckthorn-A Resource for Health and Environment in Twenty First Century* (V. Singh and P.K. Khosla, Eds.). Proceedings of International Workshop on Seabuckthorn, February 18-21, 2001, New Delhi, pp.7-13.

78. Singh, V. 2003. Geographical adaptation and distribution of seabuckthorn (*Hippophae* L.) resources. In: *A Multipurpose Wonder Plant-Botany.* Vol. I. *Botany, Harvesting and Processing Technologies* (V.Singh, Editor in Chief), p.21-34, Indus Publishing Company, New Delhi, 518p.

79. Singh, V. (Ed. in Chief). 2003. *Seabuckthorn-A Multipurpose Wonder Plant-Vol. I. Botany, Harvesting and Processing Technologies.* Indus Publishing, 518p.

80. Singh, V. 2006. Free Radicals, diseases, anti-oxidants and anti-oxidant properties of seabuckthorn (*Hippophae rhamnoides* L.). In: *Seabuckthorn-A Multipurpose Wonder Plant.* Vol. II. *Biochemistry and Pharamacology* (V.Singh, Ed. in Chief), p.3-69, Daya Publishing House, New Delhi, 600 p.

81. Singh, V. 2006. Effects of mulch, soil media and hormone on the rooting rates of hard wood cuttings of seabuckthorn. *Indian Forester.* (Submitted)

82. Singh, V., Singh, B. and Awasthi, C.P. 1995. Distribution, taxonomy and nutritional values of seabuckthorn growing in dry temperate Himalayas. In: *Proceeding of International Workshop on Seabuckthorn*, p. 52-59, Dec. 12-17, 1995, Beijing, China, 206p.

83. Singh, V. and Dogra, K.K. 1996. Characteristics, distribution, biomass, regeneration and nutritional values of seabuckthorn. *Indian Forester* 122 (6): 486-491.

84. Singh, V., Singh, R.K. and Singh, B. 1999. Fodder values of foliage of seabuckthorn (*Hippopahe rhamnoides* ssp. *turkestanica*)

growing in Lahaul valley. In: *Proc. of International Workshop on Seabuckthorn*, p. 55-59, August 29-September 2, 1999, China Science and Technology Press, Beijing, China, 97p.

85. Singh, V., Singh, R.K. and Awasthi, C.P. 2001. Morpho-biochemical changes during ripening of the fruit of seabuckthorn (*Hippophae rhamnoides* L.). *J. Non-Timber Forest Products* 8 (1/2): 114-116.

86. Singh, V. and Singh, R.K. 2004. Evaluation studies on the exotic and local genotypes of seabuckthorn (*Hippophae rhamnoides* L.). *J. Non-Timber Forest Products* 11 (3): 161-165.

87. Singh, V., Nayyar, H., Gupta, R.K. and Uppal, R. 2005. Germination behaviour of different biotypes of seabuckthorn (*Hippophae* L.). *J. Non-Timber Forest Products* 12 (4): 209-213.

88. Singh,V., Gupta, R.K., Tandon, S. and Sawhney, R.C. 2004. Fatty acid composition of fruit pulp and seeds oils of some selected forms of seabuckthorn (*Hippophae* L.). *Global Seabuckthorn Research and Development* 3 (2): 1-8.

89. Singh, V. and Gupta, R.K. 2007. Study on micropropagation of seabuckthorn. In: Int. Seabuckthorn Asso. Conf., 12-16 Aug. 2007, Quebec, Canada.

90. Singh,V., Gupta, R.K., Sawhney, R.C. and Arumughan, C. 2005. Tocopherols and carotenoid in fruit pulp oil of seabuckthorn growing in dry temperate Himalayas. In: Proceedings of 2nd International Seabuckthorn Association Conference on *"Seabuckthorn–A Multipurposes Plant, Solution to Nutrition, Health and Environment"* August 26-29, 2005, Beijing, China.

91. Singh, V. and Singh, R.K. 2004. Morpho-biochemical variations in seabuckthorn (*Hippophae* L.) populations growing in Lahaul valley, dry temperate Himalayas. *Indian Forester* 130 (6): 663-672.

92. State Standards Acting in the Russian Federation and CIS: TU 64-4-87-89: *Fresh Seabuckthorn Fruits*.

93. Tsar'kova, T.F. 1988. The effect of physiologically active substances on the rooting and growth of lignified cuttings of of *Hippophae rhamnoides*. *Agrotekhnika Selektsiya i Mekhanizatsiya v Yagodovodstve Nechernozem'ya* p: 41-46. (In Russian)

94. Tong, J., Zhang, C., Zhao, Z., Yang, Y. and Tian, K. 1989. The determination of the physico-chemical constants and sixteen mineral elements in seabuckthorn raw juice. In: *Proc. of Int. Symp. Seabuckthorn*, Xian, China, pp. 132-137.

95. Tyagi, S.P., Varshney, A.C., Kumar, A. and Singh, V. 2005. Therapeutic and prophylactic efficiencies of seabuckthorn in gastric erosion and ulcerations in dogs. In: *Seabuckthorn-A Multipurpose Plant, Solution to Nutrition, Health and Environment*. The 2nd International Seabuckthorn Association's Conference, August 26-29, 2005, Beijing, China, p. 52. (Abstract)

96. Tyurikov, E.A. 1993. Reaction of seabuckthorn to nitrogen and potassium fertilizers. In: *Fruit growing in Nonchernozemye*, p.65-69.

97. Varshney, A.C. and Tyagi, S.P. 2005. Seabuckthorn: A resource for soft tissue repair in animals. In: *Seabuckthorn-A Multipurpose Plant, Solution to Nutrition, Health and Environment*, August 26-29, 2005, Beijing, China, p. 63-64. (Abstract)

98. Wei, C. Y. and Guo, W. Y. 1996. A preliminary study on the changing pattern of oil content *Hippophaë rhamnoides* ssp. *sinensis* in Shanxi province. *Hippophaë* 9 (2): 37-43.

99. Wei, C. Y. and Guo, W. Y. 1996. A preliminary study on the changing pattern of oil content *Hippophaë rhamnoides* ssp. *sinensis* in Shanxi province. *Hippophaë* 9 (2): 37-43.

100. Xinxue, H. 1992. A study on high yield cultivation technology of seabuckthorn. *Hippophae* 3: 1-6.

101. Wolf, D. and Wegert, F. 1993. Experience gained in the harvesting and utilization of seabuckthorn, p. 23-29. In: *Cultivation and utilization of wild fruit crops*. Bernhard Thalacker Veriag Gmbh and Co., Germany.

102. Yang, B. and Kallio, H. 2006a. Lipophilic components of seabuckthorn (*Hippophae rhamnoides* L.) seeds and berries. In: *Seabuckthorn-A Multipurpose Wonder Plant. Vol. II. Biochemistry and Pharamacology* (Singh, V., Ed. in Chief). Daya Publ., New Delhi. pp. 70-97.

103. Yang, B. and Kallio, H. 2006b. Physiological effects of seabuckthorn (*Hippophaë rhamnoides*) fruit pulp and seed oils.

In: Singh, V. (Ed. in Chief). 2006. *Seabuckthorn-A Multipurpose Wonder Plant-* Vol. II. *Biochemistry and Pharamacology* (V. Singh, Editor in Chief). Daya Publishing House, New Delhi, pp. 363-189.

104. Yongshan, L., Xuelin, C. and Hong, L. 2003. Taxonomy of seabuckthorn (*Hippophae* L.). In: *Seabuckthorn-A Multipurpose Wonder Plant*. Vol. I. Indus Pub., N.Delhi, 518p.

105. Zhang, W., Yan, J., Duo, J., Ren, B. and Guo, J. 1989. Preliminary study of biochemical constituents of berry of seabuckthorn growing in Shanxi province and their changing trend. In: *Proc. of Int. Symp. Seabuckthorn*, Xian, China, pp. 96-105.

106. Zhengpin, J., Xiufeng, W. and Shunguang, L. 2007. Seabuckthorn breeding for ecological and economic improvement. In: *Seabuckthorn-A Multipurpose Wonder Plant. Vol. III.* (Singh, V., Ed. in Chief), Daya Publ., New Delhi (In press).

107. Zhijia, Y. and Xuebin, L., 2001. *China Soil and Water Conservation* (3).

Index

M

N

O

P

S